REF 001.9 STR

STRANGE BUT TRUE
9781603200318 $29.95 C2008

MAR   2008

MARILYN BRIDGES

STRANGE
BUT
TRUE

LARRY BURROWS/LIFE

DALLAS AND JOHN HEATON/STOCK CONNECTION

STRANGE
BUT
TRUE

THE WORLD'S WEIRDEST WONDERS

LIFE Books

EDITOR Robert Sullivan
DIRECTOR OF PHOTOGRAPHY Barbara Baker Burrows
CREATIVE DIRECTOR Mimi Park
DEPUTY PICTURE EDITOR Christina Lieberman
WRITER-REPORTER Hildegard Anderson
COPY Barbara Gogan (Chief), Parlan McGaw
PHOTO ASSISTANT Forrester Hambrecht
CONSULTING PICTURE EDITORS
Mimi Murphy (Rome), Tala Skari (Paris)

PRESIDENT Andrew Blau
BUSINESS MANAGER Roger Adler
BUSINESS DEVELOPMENT MANAGER Jeff Burak

EDITORIAL OPERATIONS Richard K. Prue,
David Sloan (Directors), Richard Shaffer (Group Manager),
Brian Fellows, Raphael Joa, Angel Mass, Stanley E. Moyse,
Albert Rufino (Managers), Soheila Asayesh, Keith Aurelio,
Trang Ba Chuong, Charlotte Coco, Osmar Escalona,
Kevin Hart, Norma Jones, Mert Kerimoglu, Rosalie Khan,
Marco Lau, Po Fung Ng, Rudi Papiri, Barry Pribula,
Carina A. Rosario, Christopher Scala, Diana Suryakusuma,
Vaune Trachtman, Paul Tupay, Lionel Vargas, David Weiner

TIME INC. HOME ENTERTAINMENT
PUBLISHER Richard Fraiman
GENERAL MANAGER Steven Sandonato
EXECUTIVE DIRECTOR, MARKETING SERVICES Carol Pittard
DIRECTOR, RETAIL & SPECIAL SALES Tom Mifsud
DIRECTOR, NEW PRODUCT DEVELOPMENT Peter Harper
ASSISTANT DIRECTOR, BRAND MARKETING Laura Adam
ASSOCIATE COUNSEL Helen Wan
SENIOR BRAND MANAGER TWRS/M Holly Oakes
BOOK PRODUCTION MANAGER Suzanne Janso
DESIGN & PREPRESS MANAGER Anne-Michelle Gallero
SENIOR MARKETING MANAGER Joy Butts
BRAND MANAGER Shelley Rescober

Special thanks to Alexandra Bliss, Glenn Buonocore,
Susan Chodakiewicz, Margaret Hess,
Arnold Horton, Robert Marasco, Dennis Marcel,
Brooke Reger, Mary Sarro-Waite, Ilene Schreider,
Adriana Tierno, Alex Voznesenskiy

Copyright 2008 Time Inc. Home Entertainment

Published by
LIFE Books

Time Inc.
1271 Avenue of the Americas
New York, NY 10020

All rights reserved. No part of this book may be reproduced
in any form or by any electronic or mechanical means,
including information storage and retrieval systems,
without permission in writing from the publisher, except by
a reviewer, who may quote brief passages in a review.

ISBN 10: 1-60320-031-2
ISBN 13: 978-1-60320-031-8
Library of Congress #: 2008903294

"LIFE" is a trademark of Time Inc.

We welcome your comments and suggestions
about LIFE Books. Please write to us at:
LIFE Books
Attention: Book Editors
PO Box 11016
Des Moines, IA 50336-1016

If you would like to order any of our
hardcover Collector's Edition books,
please call us at 1-800-327-6388
(Monday through Friday, 7:00 a.m.–8:00 p.m.,
or Saturday, 7:00 a.m.–6:00 p.m., Central Time).

Driftwood Library
of Lincoln City
801 S.W. [   ] 01, #201
Lincoln City, Oregon 97367

CONTENTS

6 INTRODUCTION
"True" Is in the Mind of the Beholder

8 STRANGE BEINGS

40 STRANGE BUT NOT TRUE

44 STRANGE DOINGS

78 STRANGE BUT NOT TRUE

82 STRANGE PLACES

124 STRANGE BUT NOT TRUE

128 JUST ONE MORE

Endpapers: Nazca Lines (please see page 88)

"TRUE" IS IN THE MIND OF THE BEHOLDER

When we at LIFE Books approach a subject—be it, to cite recent examples, violent weather, the civil rights movement or the turbulent life of Frank Sinatra—we do so with the conviction that we will put before you, our readers, the truth, the whole truth and nothing but the truth. When we elect to produce a book called *Strange But True,* which will deal with all manner of weird and wondrous phenomena in the natural and even the unnatural world, we cross our fingers. There are two reasons: For good luck, as we relate some admittedly dicey tales. And in case—just in case—we let drop the occasional fib, not to say the true whopper, along the way.

ELSIE WRIGHT

You see, "facts" that might seem like real stretchers attach themselves to the narratives of such as the Loch Ness Monster, Bigfoot and the stone circle at Stonehenge like so many barnacles, and they just can't be pried off. What we've attempted to do in our book is keep the narrative lively and fun but call a spade a spade as often as we can. We'll tell you what is known about these subjects, and what is conjectured. When dealing with far-out occurrences in places from the Tunguska Forest in Siberia to the ranch outside Roswell, New Mexico, we may seem to some, upon reading our speculations, too gullible by half. But true believers believe these things to be true, and our job, as we see it, is to give you the relevant theories, to explain the possibilities and impossibilities, and then let you decide.

This introduction is by no means a disclaimer. Folks far smarter than us firmly believe that some small percentage of UFO sightings have involved extraterrestrial beings, as you will learn in the Roswell section of our book. The world's preeminent mountain climber has testified that he has seen the Abominable Snowman—or, at least, the Himalayan creature that passes for the Snowman. The hunt for buried treasure on Canada's Oak Island is more than two centuries old, and continues today. Many, many people take this stuff very, very seriously.

And others just have fun with it. We hope what you learn in these pages will be interesting to you, and as a bonus that you'll take away some really neat water-cooler–conversation starters. Did you know that an agency of your United States government has certified a house in California as truly haunted? It has. Here's another: The same man was hit by lightning seven times and lived to tell the tale. A few more: The biggest-ever animal, much larger than any dinosaur, is still among us. And it's actually true that no two snowflakes are alike, but it's not true that the Komodo dragon is a dragon (although it looks just like one).

You will learn what we know, what we suspect and what we do not know about the yeti, about the whale with the horn of a unicorn, about the statues on Easter Island, about the world's deepest cave and its tallest man, about airborne mammals (bats do fly, but flying squirrels—not to mention reindeer—do not).

You will meet compelling people in these pages, from Vladimir Lenin to Thomas Jefferson and John Adams to the guy who would be Dracula. You will also learn in our "Strange But Not True" chapters about things that have been incontrovertibly proven to be false. Sorry, but the Cottingley fairies were made out of cardboard, and that gigantic mutant cat you saw on the Internet is in real life a normal-size kitty. At the risk of bringing the wrath of all *Da Vinci Code* disciples down upon us, we lay out here the facts, which are verifiable, behind the bogus Priory of Sion.

Now, the name of our book is *Strange But True,* and we can say with confidence: If some of the stories in this book may, in fact, one day be proved to be untrue, none of them can ever be said to be not strange. Even perfectly natural phenomena like the northern lights, which can be explained as clinically as can wind or any other meteorological happening, are strange and wondrous. They are capable in their ethereality of spurring myth just as easily as they are of inspiring awe.

There is oddness here, there is spirited fun and there is occasional beauty. There is, we guarantee, never a dull moment. The strangeness of the world around us is fascinating in the extreme.

And that's the truth.

● **Opposite:** On remote Easter Island in the Pacific, the famous statuary stands watch by night. **Left:** Frances Griffiths poses amidst the Cottingley fairies for her friend Elsie Wright's camera in 1917. **Below:** The world's only flying mammal, the bat, hanging out. **Bottom:** The aurora borealis colors the sky above Alaska.

BOB ELSDALE/AURORA

MICHIO HOSHINO/MINDEN

STEPHEN ALVAREZ/NATIONAL GEOGRAPHIC

STRANGE BEINGS

GETTY

We begin our histories of things that are either demonstrably true (the great majority of them) or false (who can say?) with one of the most famous and controversial tales ever told: that of the Loch Ness Monster. A millennium and a half after first reports of a large reptilian creature trolling the lake were issued, skeptics still ask "Where's the proof?" while adherents point to the thousands of claimed sightings.

Loch Ness is a large, narrow body of water extending 24 miles in length and having a surface area of 13,952 acres. Nestled amidst the rugged mountains of the Northern Highlands of Scotland, it is extremely deep: Much of it has twice the depth of the North Sea, and the bottommost point of the loch is some 750 feet below the wind-tossed waves. Fed by seven major rivers, it contains more water than all the other lakes in England, Scotland and Wales put together. Remote, in a fierce landscape, it would indeed be a fine place for a monster to dwell.

P.A. MACNAB

FISHING FOR NESSIE

The famous Wilson photo of 1934 (above) was later revealed to be a fake, but there have been many other intriguing pictures, including this one from 1955 that shows a large object in the water near Urquhart Castle.

In the late 1990s, divers used cow's blood and a bait basket trolling 60 feet below the surface.

SIRBIS
CORBIS

● In June 2003, the white witch Kevin Carlyon used his own, less high-tech methods to summon Nessie.

EMORY KRISTOF/NATIONAL GEOGRAPHIC

And perhaps it is that. As long ago as the 6th century, Saint Adamnan wrote an account of Saint Columba, a former Abbot of Iona, "driving away of a certain water monster by virtue of prayer of the holy man. At another time, again when the blessed man was staying for some days in the province of the Picts, he found it necessary to cross the river Ness; and when he came to the bank thereof, he sees some of the inhabitants burying a poor unfortunate man whom, as those who were burying him themselves reported, some water monster had, a little before, snatched at as he was swimming and bitten with a most savage bite."

Through the ages, other testimony accrued. Not all versions of Nessie looked alike; some were more serpentine than others, some resembled a horselike creature, some were said to be 20 feet long and others were twice that length. Of course, these discrepancies, along with sightings of multiple monsters, might indicate a breeding colony, which would certainly be necessary for the tale to be true—otherwise, Saint Columba's monster would be impossibly old in the present day, a freshwater Methuselah.

A popular conception of the creature has described it as a marine reptile with a large body, flippers, a long neck and a small head, not dissimilar to a long extinct dinosaur, the plesiosaur. This image was reinforced mightily in 1934 when a Dr. Robert Kenneth Wilson produced a photograph of just such a beast and alleged that it had been snapped at Loch Ness. For decades, the Wilson photo represented the face of Nessie. Then in the 1990s, researchers David Martin and Alastair Boyd, who had embarked on a scientific mission to find the monster, revealed the photograph as a fake. An alleged deathbed confession by a man who had been involved in the prank many years earlier said the picture actually showed a small sea serpent made of plastic wood attached to a toy submarine.

And yet there still came, through the years, many more photographs of Nessie—none of them conclusive but some more persuasive than others—and added to these, a host of additional claims of sightings. None other than Alastair Boyd, who helped unmask the hoax behind the Wilson photograph, claimed he himself had seen a whalelike creature approximately 20 feet long in the loch. Researchers who couldn't buy into the remnant-plesiosaur theory put forth new postulations: Maybe Nessie was a sturgeon on steroids?

In 2003, the BBC produced *Searching for the Loch Ness Monster,* which involved, in part, a canvassing of the entire loch using 600 sonar beams issuing from satellites. Said Ian Florence, who helped in the BBC's hunt: "We went from shoreline to shoreline, top to bottom on this one. We have covered everything in this loch and we saw no signs of any large living animal in the loch."

Case closed? Certainly not. Last year there were sightings, and next year, no doubt, there will be sightings. Nessie pilgrims will continue to venture to Scotland, questing after the life-changing glimpse, perhaps the history-making picture. And lest you think that it is only the romantics or the nutcases who refuse to let Nessie rest in peace, consider this summation from a current edition of the super-serious *Encyclopedia Smithsonian:* "To date, the actual existence of a monster in Loch Ness has not been proven. Even though most scientists believe the likelihood of a monster is small, they keep an open mind as scientists should and wait for concrete proof in the form of skeletal evidence or the actual capture of such a creature."

Fishing season remains open for Nessie, the world's most cherished monster.

KLEIN/PETER ARNOLD

Think of Australia as an ark. Eons ago, when the island continent drifted away from the landmass that would become South America and Africa, it went sailing solo, its fate its own. This was very good news for several animal species aboard the Good Ship Oz. They would enjoy relative peace through the millennia, as not all of the predators that were then roaming the larger continents made the leap.

Because Australia is in a unique, isolated situation, many of its animals appear to be singular and strange to our eyes. Hopping kangaroos, stilt-walking emus and gum-chewing koalas look odd to us, even as they are business as usual for Aussies.

ODDITIES DOWN UNDER

The platypus is certifiably bizarre no matter the audience. It is a mammal, but one apart from all others. Beginning with the front end, there is that snout, which looks like a duck's bill, though it is in fact a protuberance of soft, leathery skin. The legs of the platypus extend outward like a lizard's, and the feet are webbed, which is invaluable because the platypus spends much of its time under water searching for worms, shrimp and insects. Below the surface it closes its eyes and relies on receptors in its snout to locate the electrical field emitted by the muscles of its prey. The platypus may dive as many as 80 times in an hour, but it can stay down for 10 minutes by remaining inactive.

As with all mammals, the mother feeds her young (usually two) with her milk, but she has no teats. Instead, the milk is secreted through tiny pores in her belly, coating the fur so that the young can suck it up. (Just by the way, these young were hatched from eggs, yet another resemblance to reptiles.) As for the males, on the inside of each hind paw, they carry a little spur that can inflict a venomous sting capable of causing excruciating pain to people and potent enough to kill a dog.

The platypus is a weird, wonderful thing, neither this nor that—yet wholly itself.

The Tasmanian devil is certainly weird as well: a solitary, nocturnal, carnivorous marsupial that displays raucous anger when threatened and equally demonstrative—and loud—glee when eating. Its screech is bloodcurdling, its appetite voracious. If Australia can be thought of as an ark, then Tasmania, the island off its southern shore, is a lifeboat for this creature, which was extirpated from the mainland some 600 years ago. Today, in its heretofore safe homeland, the devil is threatened by a new disease, devil facial tumor, and in May of 2008 was declared endangered.

DAVE WATTS/PETER ARNOLD

● The platypus appears serene as it swims near the surface in its native Australia; the Tasmanian devil seems much less so. But appearances can deceive: Were an animal the size of Tazzie to get stung by a male platypus, he'd likely be a goner.

GLEB GARANICH/CORBIS

AP

● Above: It's not at all in the genes: Stadnyk dwarfs his mother, as Wadlow did his family (left).

● Opposite, far right: A "hobbit" skull from Flores is far smaller than that of a standard member of Homo sapiens. Right: This "Loulan Beauty" mummy, some 4,000 years old was a surprising find in China.

T

There is, in truth, no such thing as "the man in the street." Each of us is a unique individual with personal traits, habits and characteristics. Some of us stand out in subtle ways, some in ways that are more overt.

The Ukrainian Leonid Ivanovych Stadnyk stands out extraordinarily. At 8 feet 5.5 inches (as of 2006), the former veterinarian is the world's tallest living man. Stadnyk, born in 1971, had brain surgery at age 14 and subsequently developed a tumor that caused his pituitary gland to secrete excessive amounts of growth hormone, stimulating his spurt to enormous size (he weighs more than 400 pounds). Because he is so big, transportation, public or private, is a problem, and he stays largely in his village 130 miles from Kiev. "My height is God's punishment," he has said. "My life has no sense."

THE HUMAN CONDITION

Stadnyk is still growing and could one day become the tallest man in medical history, a designation now held by the late Robert Pershing Wadlow, who was born in 1918 in Alton, Illinois, and died of an infection at age 22. Wadlow, too, had an overactive pituitary gland that spurred his growth to a height of 8 feet 11.1 inches.

There are many other examples of human anomaly, some discovered only after thousands of years have passed. Archaeologists were startled in the early 1990s, for instance, to discover the mummies of Indo-Europeans in China's forbidding Takla Makan Desert, indicating that a previously unknown civilization had flourished along the Silk Road some 4,000 years ago. And in 2003, on the Indonesian island of Flores, a near-complete skeleton of a three-foot-tall adult woman with a small skull led to speculation about a new human species. Whether the so-called "hobbits" of Flores were in fact a separate species living alongside homo sapiens 18,000 years ago is still being debated today, especially in light of other tiny skeletons that were discovered on the nearby islands of Palau in 2008. This later find implies that the hobbits may have been what are called insular dwarfs: regular humans who were undersized due to their restricted life on islands.

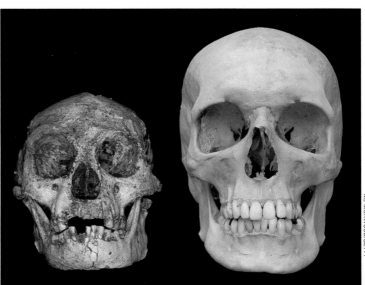

REZA/NATIONAL GEOGRAPHIC

M.J. MORWOOD/GETTY

The world said hello to Dolly on February 22, 1997. But when her birth was announced on that day, not everyone reacted with applause. This was because the ewe was a clone, the first animal to result from a cell taken from an adult of its species that was stimulated and then implanted in a surrogate mother. In Dolly's case, the cell had been part of a mammary gland, and the lamb was named after the cheerfully busty queen of country music, Dolly Parton.

CLONING AROUND

Following the lead of the Scottish scientists who succeeded with Dolly, others around the world proceeded to replicate mice, cows, horses and even dogs. Boosters of the revolutionary procedure saw it as a possible way to one day preserve endangered species or to augment the food supply. Critics took various different tacks. Some argued that cloning animals was ethically wrong, against nature, and that it moved us perilously close to human cloning, which they viewed as an abomination before God. Others pointed out that cloning today was an inexact science resulting in many failures and deformed animals, and that, besides, it could not avoid longstanding problems associated with inbreeding.

Nevertheless, there are now cloned animals among us, something that was a science fiction notion only a short while ago. Sadly, Dolly is no longer one of their number. She lived for six years at the Roslin Institute in Edinburgh, where she was born, producing four lambs of her own during that time. After developing a progressive lung disease, she was euthanized.

Even in death, Dolly has fed the controversies that had been ignited by her birth. Her keepers said the cancer she had developed was common in her breed and that there was no seeming connection to her being a clone. But others noted that Finn Dorset sheep such as Dolly have a life expectancy of a dozen years or more, and wondered, since she had been cloned from an animal that was already 6 years old, whether half her life was behind her even as she was being weaned.

REMI BENALI/CORBIS

R. BACH/CORBIS

RICHARD OLSENIUS/REUTERS

● Dolly, the Adam and Eve of cloning (opposite, with her godfather, Scottish scientist Dr. Ian Wilmut), was followed by cats and monkeys (above) and dogs such as Snuppy (left), who was born in South Korea from a cell taken from this larger Afghan hound.

SEOUL NATIONAL UNIVERSITY/CORBIS

In mythology, the dreaded Hydra was a beast with a serpent's body and multiple heads that was slain, finally, by Hercules. In Byzantine heraldry, the double-headed eagle symbolized the emperor's dominance over society's secular and religious arms.

In reality, bicephalic, or two-headed, cattle roam the range and bicephalic snakes slither through the grass. These anomalies are rare, to be sure, but they do exist and perhaps informed the legendary polycephalic monsters of yore.

REPTILEAN ANOMALIES

Such animals are the products of the same process that results in conjoined human twins: the union of two embryonic disks. Life presents complexities for all two-headed animals but, it has been observed, particular problems arise for the snake because of its single-minded focus on its own survival. As each head has its own brain, there is often disagreement on where to go or what to do: If attacked, the reptile has trouble responding quickly; and if prey is at hand, the heads will vie for which one gets to eat. Sometimes the two heads fight, and if one of them senses food on the other, it might try to devour it. For all of these reasons, two-headed snakes are ill-suited to life in the wild. But some that have been captured have fared well. Thelma and Louise, a two-headed corn snake, lived at the San Diego Zoo for a number of years. We, a two-headed albino rat snake, lived for eight years at the World Aquarium at St. Louis's City Museum.

Other reptiles that give pause—and there are more than a few—would certainly include the curious Jackson's chameleon. This African lizard is also known as the three-horned chameleon because the male of the species looks like a foot-long triceratops, with a horn on its nose and one by each eye. And get this: Each of those eyes is not only independent of the other but is able to rotate 180 degrees. Nature doles out some nifty tricks indeed.

KIM TAYLOR/MINDEN

DANI CARDONA/CORBIS

● The sticky tongue of the three-horned chameleon is quite a weapon, as a fly learns. This two-headed false smooth snake was found in a hotel garden on Palma de Mallorca, a Mediterranean island, in 2002.

Few things appear scarier than giant reptiles or rodents, and here we meet a few of the world's most gargantuan. The reticulated python of Southeast Asia is the world's longest snake, at 30 feet or more in length—longer even than a boa constrictor or anaconda. Like a boa, the python suffocates its prey, which can include animals as big as a dog or a pig, through constriction.

BIG FELLAS

More disturbing than that is the killing method of the world's largest lizard, the Komodo dragon, which roams the Indonesian islands of Komodo, Gila Motang, Rinca and Flores. As is the python, this fearsome nine-foot-long carnivore is an ambush hunter, lying in wait for prey to pass by. Then it springs (and if it needs to, sprints in short bursts at up to 11 miles per hour). It tears into its target with claws and with teeth that are serrated like a shark's. This king of all monitor lizards proceeds to disembowel its unfortunate victim and scarf down every part of it. The Komodo dragon can eat up to 80 percent of its own weight, which means that a 200-pound male can ingest in one meal more than 150 pounds of water buffalo, boar or any other delicacy, including juveniles of its own species.

GARY VESTAL/GETTY

Disgusting on a different scale, at least as judged by human standards, are the dining habits of the world's largest living rodent, the South American capybara. Its name derives from a term meaning "master of the grasses" in the Guarani language. And indeed, the diet of this semiaquatic herbivore consists largely of grass (six to eight pounds a day), plants, fruits and tree bark. In order to aid in the digestion of the vegetation's cellulose, capybaras eat their own feces for its bacterial component. This dietary regime is enough to sustain a stocky animal that grows to more than four feet in length and 140 pounds in weight. But consider this: In ages past, there existed a now extinct capybara that was eight times as large—larger than a grizzly bear.

The capybara, we hasten to note, is a gentle giant. Sometimes called "the world's biggest rat," it is actually a much closer cousin to a guinea pig, and very much likes to be petted.

THEO ALLOFS/GETTY

RUSSELL MCPHEDRAN/AP

THEO ALLOFS/GETTY

A sense of scale is provided for a 15-foot-long python by five keepers at the Australian Reptile Park (opposite, bottom), and for a capybara in Brazil by a little bird (left). Below: Two Komodo dragons prowl a beach on their namesake island in a scene that seems lifted from *Jurassic Park.* A Komodo's claws (opposite, top) are as efficient as they are fearsome.

T

The natives of far-northern Alaska call the musk ox "oom-ingmak," meaning "the animal with skin like a beard." Of course, it is the musk ox's wool, not its skin, that is shaggy (and is said to be the warmest in the world, which is what protects the animal at temperatures that fall to 70 below), but then again: The English name, musk ox, is also misleading, for the animal is quite clean in the wild and does not smell at all musky. This mammoth mammal is kin to sheep and goats, believe it or not, though the male grows to a height of up to five feet at the shoulder and weighs in at up to 800 pounds.

THROUGH THE MISTS OF TIME

SARAH RICE/CORBIS

It looks primeval and in fact has changed little since the Ice Age. Some of its instincts, too, seem pulled from the past. When threatened by wolves or grizzlies, the adults form a circle around their calves, looking outward. This is an effective defense. If a predator attacks, a butt from a severe musk ox head usually settles the matter. Four inches of horn protrude from the male head and slightly less from the female. The males have three more inches of bone on the forehead to protect the cranium. Analysis of musk oxen head-butting during mating season after charging 50 yards at one another puts the force of the impact at that of a car ramming a concrete wall at 17 miles an hour.

Eons ago, and still in certain northern regions today, most horses were quite small relative to what we regard as the equine norm. Their size was largely determined by the availability of food and by climate; a stockier horse fared (and fares) better in Iceland, for instance, than a strapping stallion might. The horse seen here is not necessarily representative of these hearty subspecies of yore, but is the product of a modern practice: the purposeful breeding of dwarf horses. Thumbelina, currently the world's smallest living horse, was born in 2001 on Goose Creek Farm in St. Louis to standard miniature-horse parents. Regular miniature horses grow to be about three feet tall, but Thumbelina topped out at 17.5 inches and close to 60 pounds. The little lass is a favorite of all children who get to stroke her.

● Thumbelina seems custom-built to delight children, such as these at the Children's Specialized Hospital in Fanwood, New Jersey, in 2007. Musk oxen are customized, too: for a harsh environment. Below: Musk oxen with their calves.

E. & P. BAUER/CORBIS

W

Where to begin with the octopus? An invertebrate—no bones—it is basically just a head with arms (eight of them, as any schoolkid knows). Its three hearts, pumping the mysterious mollusk's light-blue blood, are located in the head, as are the stomach and all the other vital organs. Did you know that in its mouth is a beak, hard as a parrot's? And that the mouth is completely encircled by those flexible arms with the suckers on them, which are used to grab and sample foods? Into the beak go preferred treats such as clams and crab or lobster meat.

DENIZENS OF THE DEEP

Of the more than 150 types of octopus inhabiting mostly tropical and subtropical waters, some are only about half an inch across, while the giant Pacific octopus can grow to 30 feet and 600 pounds. When threatened, an octopus can change color as a means of camouflage or aggression by deploying pigment cells attached to its muscles and controlled by its nervous system: An octopus that has turned white is afraid; one that's turned red is angry. (Other possibilities are blue, gray, purple, brown and, yes, striped.) Under extreme duress, it excretes a poisonous ink and darts away behind the obscuring cloud. The common octopus is considered the smartest of all invertebrates.

Even more bizarre than the octopus are the denizens of a dark world extending down from 1,000 to 27,000 feet, the mesopelagic species and bottom-feeding benthic fishes—anglerfish, swallowers, eels and a creature known in the oceanographic world as "the vampire squid from hell." They resemble ancient animals rather than anything else alive today, and possess—variously—radar, ferocious dental work and a supernatural talent for bioluminescence. The members of this deepwater menagerie seem to have sprung full-blown from the imaginations of Hieronymus Bosch or Tim Burton, yet they are part of our world. Pleasant dreams . . .

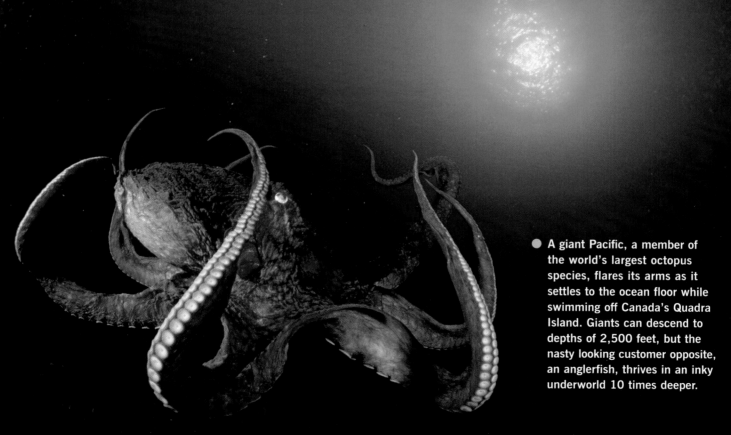

● A giant Pacific, a member of the world's largest octopus species, flares its arms as it settles to the ocean floor while swimming off Canada's Quadra Island. Giants can descend to depths of 2,500 feet, but the nasty looking customer opposite, an anglerfish, thrives in an inky underworld 10 times deeper.

FRED BAVENDAM/MINDEN

NORBERT WU

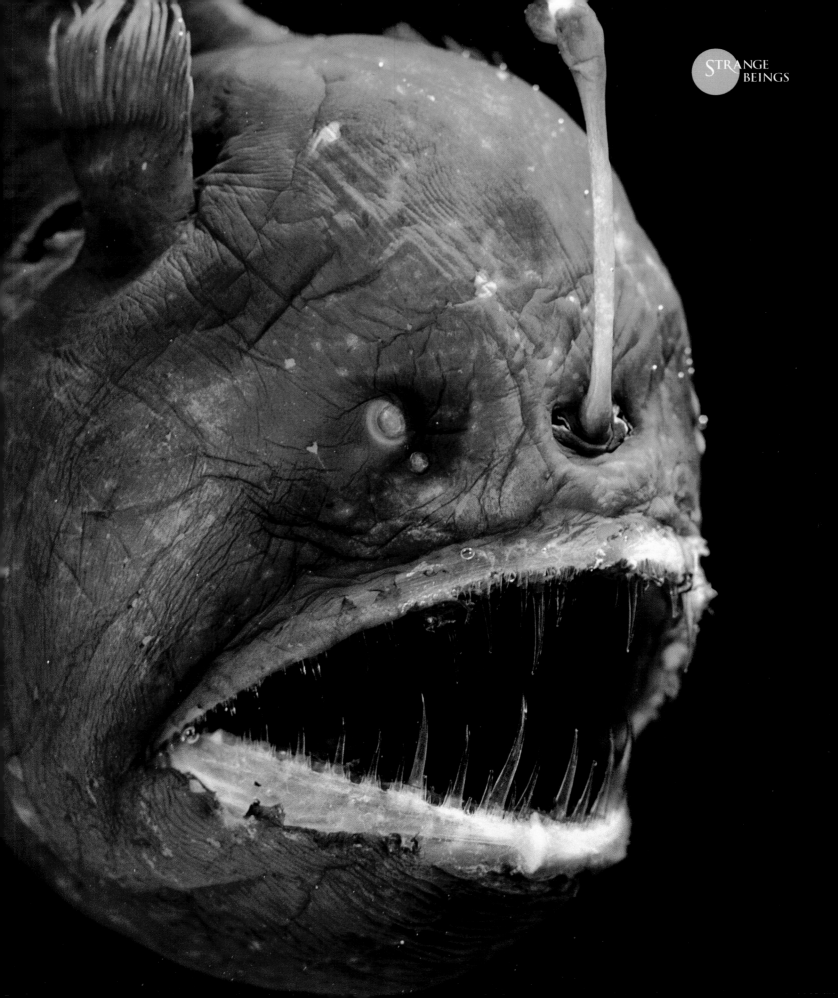
STRANGE
BEINGS

● Here are two views of blue
whales off the coast of Baja
California, southwest of San
Diego. The animal is so big,
the diver seen here in the
foreground could crawl
through its aorta (not that he
would want to). These
whales have a life span of
between 35 and 40 years.

It is fortunate that the largest animal ever to inhabit the earth—larger than the largest dinosaur would have been—is still among us. The blue whale, which certainly could never live at its size on land (the water supports its massiveness), is almost inconceivably big. It can grow to 105 feet in length and 200 tons (that's 400,000 pounds!) in weight. Metaphors and comparisons might help. Look at a Boeing 737 on the runway. That's a blue whale. Look at that Honda Civic parked over there. That's a blue whale's heart. Look out at your back patio. That's a blue whale's tongue—large enough for 50 people to stand on.

THE HEAVYWEIGHT CHAMP

A baby blue whale—at the moment of its birth—weighs two tons, and then begins drinking 50 gallons of its mother's milk each day, gaining about 200 pounds a day during its first year. This huge mammalian carnivore, once weaned, dines almost exclusively on tiny shrimplike animals called krill, which it ingests as it swims. An adult blue whale eats as much as 40 million krill—four to eight tons of krill—per day.

You might suppose krill is an endangered species, but it is not. The blue whale, however, is.

For decades the blue whale was a trophy catch of the whaling industry as it represented the potential for so much oil and blubber in a single kill. And we are not referring just to the days of Herman Melville here; in a whaling season as recent as that of 1930–31, nearly 30,000 blue whales were taken by the trade. The whale was nearly extinct when it was agreed by the International Whaling Commission in 1965 to give protection to blues and other whales. It is far too soon to characterize the blue whale's stabilization since then as a comeback, but today there are estimated to be between 3,000 and 5,000 of these giants roaming the world's oceans. The planet's largest-ever citizen is, yes, still among us.

MARK JONES/NATIONAL GEOGRAPHIC

MIKE JOHNSON

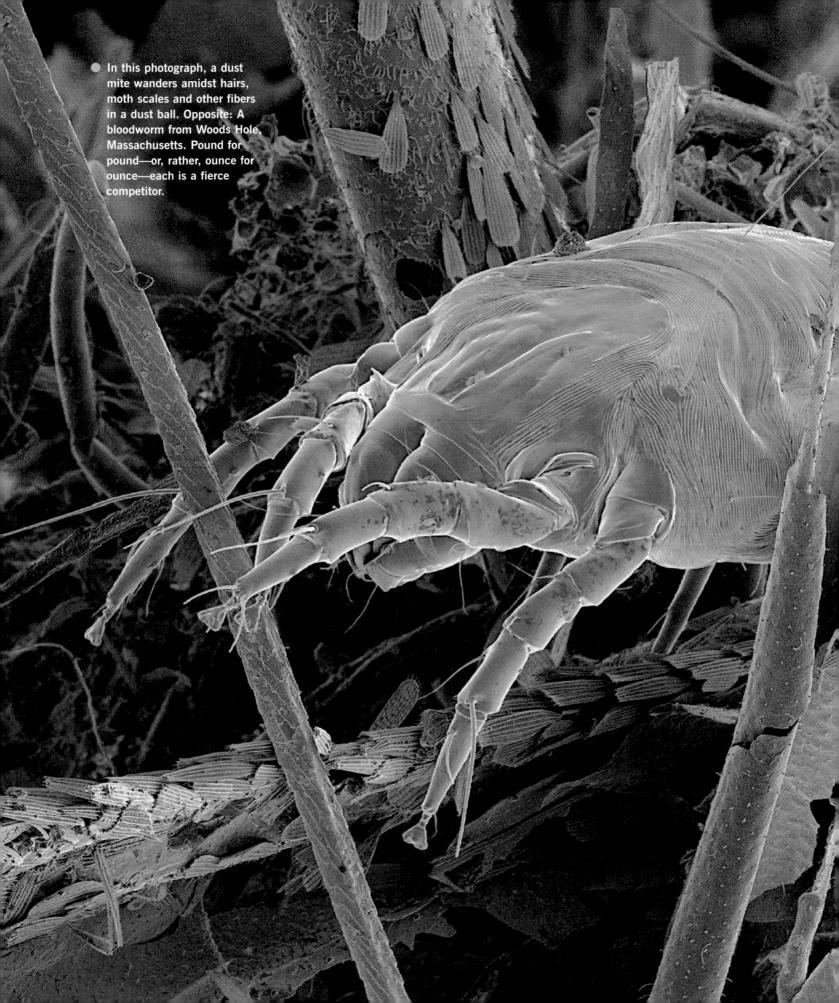

In this photograph, a dust mite wanders amidst hairs, moth scales and other fibers in a dust ball. Opposite: A bloodworm from Woods Hole, Massachusetts. Pound for pound—or, rather, ounce for ounce—each is a fierce competitor.

KJELL B. SANDVED/PHOTO RESEARCHERS

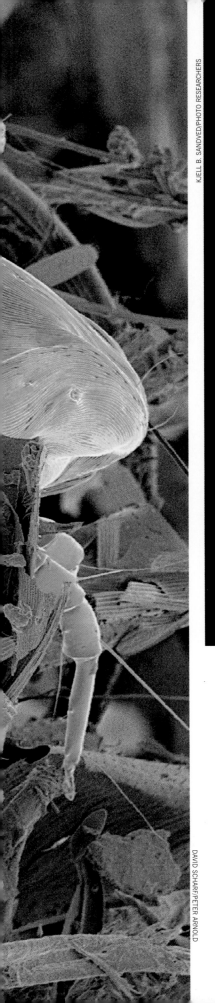

DAVID SCHARF/PETER ARNOLD

LITTLE MONSTERS

When we think of sea monsters, we might picture a giant squid as imagined by Jules Verne. As he tells us, it's an enormous thing that can menace a submarine. We shiver in terror.

Well, meet Glycera, up to 14 inches of fighting worm that, ounce for ounce, can hold its own as a battler in marine waters. Also called a bloodworm, it burrows into the sand and behaves as an "active raptorial species" or, more descriptively, an "ambush raptor." The little fellow uses a powerful pharynx that is concealed in its interior and can be expelled forcefully to either dig into sediment or to attack a foe; the worm can literally turn itself inside-out through its mouth (a talent called being "eversible"). A largish, annoyed Glycera can issue a bite to a human that stings with poison from a gland in the pharynx wall; to the small invertebrates upon which it preys, the bloodworm's kiss is lethal.

Glyceridae prove that ferocious things come in small packages. Much smaller still is a creature that lives largely by eating dead human flesh. This is . . . the dust mite.

Consider this: You spend a third of your life in bed and, perhaps unbeknownst to you, you share that mattress nightly with between 100,000 and 10 million microscopic dust mites. Their food of choice is dander, which consists of human and animal skin flakes. We each shed about a fifth of an ounce of dander per week, which constitutes a bounteous feast for our resident mites.

The good news is that dust mites are not parasites; they eat only dead tissue. The bad news is that, while they don't carry disease, they can trigger allergic reactions; and they do secrete waste—up to 20 droppings per day. Ten percent of the weight of your two-year-old pillow might well comprise dead dust mites and their excrement.

Certainly the best news, as applies to both Glyceridae and dust mites, is: out of sight, out of mind.

Until just now.

THEO ALLOFS/PETER ARNOLD

● A bat takes his rest, while (opposite) a southern flying squirrel soars through the air with the greatest of ease and an aye-aye plots his next move. He uses that extra-long finger to probe for insects, his food, in tree branches.

M Mammals don't fly—everybody knows that. Or do they? The bat is a mammal, and it flies. Doesn't it? Well, yes, in fact, it does: It is true flight that the bat is capable of. This remarkable little creature with the out-sized reputation is a catalyst for ear-piercing shrieks and vampire dreams and in fact the only mammal with the talent to fly (discounting, of course, Santa's reindeer).

KIM TAYLOR/MINDEN

ANIMALS GONE BATTY

The bat is one of the planet's most famous, notorious and ubiquitous critters. The thousand species of bats, divided into vegetarians and carnivores and ranging in wingspan from six inches (the bumblebee bat) to six feet (the flying fox), account for fully a quarter of all mammalian species on earth. They are everywhere, they are plentiful, and they are not the villains that they are often made out to be. Perhaps evolved 100 million years ago from a nonflying squirrel-like mammal that leapt from tree to tree, some bats had more skin between their arms and body which, when flapped, allowed them to rise and glide.

Bats do lots of strange things: They live in caves, they hang upside down, they source food through echolocation (bouncing a sound off a target to detect it). And they eat lots of insects, some of them up to 600 an hour, which is why they should be considered to be our friend not our foe.

The flying squirrel does not fly, not really. It has a membrane, a fold of skin, between its front and hind legs which, when the animal leaps from the trees in which it lives, allows it to glide exceptionally well, even to bank and make acrobatic turns. The flying squirrels that inhabit the eastern United States can soar approximately 50 yards, while the giant Southeast Asian species can glide more than twice that distance. The squirrels are nocturnal and have large eyes that afford them night vision when zipping about in the dark.

Another nocturnal, goggle-eyed tree-dweller is the bizarre aye-aye, which lives only off the east coast of Africa on Madagascar. The island is renowned as the "land that time forgot" due to its wacky collection of animals. The aye-aye looks like a rodent; and indeed, it has front teeth that grow continuously like a squirrel's and ears that resemble a bat's. But it is in fact a lemur, a primate, and thus related to chimpanzees, apes and . . . well, you and me. This freaky animal is supposedly an omen of bad luck, even death, and it's apparent at a glance how it assumed such a low reputation.

PETE OXFORD/MINDEN

The vast distances that some animals travel during their natural life cycles, either to escape harsh climates on a seasonal basis or for other reasons, are mind-boggling. Among the heroic migrations accomplished by animals ranging from little locusts to big caribou, two stand out.

NOTHING IF NOT MIGRATORY

Salmon are born upstream. When they reach their juvenile stage, after a year or more, they have developed some characteristics that will allow them to thrive in the ocean and have moved downriver to the estuary where freshwater and saltwater mix. There, they double or triple in size before taking the plunge and heading out to sea.

The ocean will be their home for one to five years, depending upon the species. A diet of shrimp, anchovies, herring and other small fish will make them strong in preparation for the final chapter in their story: the return home to spawn.

It is not known precisely how salmon navigation works, but it is thought that a combination of ocean currents, relative salinity, relative temperatures, guidance by the sun and the stars and even a sensitivity to the earth's magnetic field leads the fish back near the mouth of their home river. A salmon can actually detect the odor of its native stream. Some salmon do stray to the wrong one, but the accuracy of the great majority is astonishing. Many ascend to the very spot where they were born, and here they spawn the next generation.

The monarch butterfly's transformation from egg to caterpillar to chrysalis to adult is miraculous enough. That many of these insects, which weigh barely more than a hundredth of an ounce and feature a wingspan of just four inches, can accomplish a 2,000-mile trip to their ancestral wintering ground seems simply too much.

Some butterflies can overwinter in certain environments, but the monarchs that summer in Canada and the eastern United States cannot. So this tiny being has learned to migrate to southern states or to Mexico. (Some wandering monarchs, when the wind was right, have even made transatlantic crossings—a truly rare achievement for an insect.) Best guesses as to how the butterflies find their way suggest they use the sun as a compass. It is a remarkable feat by any measure, and made much more so because other monarchs, in other areas, do not migrate at all.

● Pink salmon spawn en masse in an Alaskan riverbed, while monarch butterflies congregate en (an even bigger) masse at a tree in Michoacán, Mexico. Both of these photographs represent journey's end.

MICHAEL QUINTON/MINDEN

FRANS LANTING/AURORA

STRANGE
BEINGS

BIRDS NOT OF A FEATHER

● Here, a wandering albatross gives an elaborate courtship performance for a potential mate on South Georgia Island in the southern Atlantic. Opposite: A female North American ruby-throated hummingbird in flight.

H

ummingbirds are unique characters in many ways. Flapping their wings from 15 to 80 times per second, they not only create their distinctive humming sound but they are able to hover in midair. They have such aerodynamic control, in fact, they can fly backward—the only birds that can turn that trick—as well as vertically, laterally and upside down. They can accelerate from zero to 25 miles per hour in no time and sometimes exceed twice that speed during courtship dives (at which point their wings are frantically flapping at up to 200 times per second). This tiny little superstar is an avian athlete of the highest order. It's all about energy.

Hummingbirds have a high breathing rate, heart rate (up to 1,260 beats per minute in the blue-throated hummingbird) and body temperature—and, while in flight, the highest rate of metabolism of any animal on the planet save insects. In order to zoom about as they do, they need lots of fuel, and a hummingbird consumes one half or more of its body weight in nectar each day. As you might imagine, that means regular feedings are mandatory, and the birds visit flowers or man-made feeders between five and eight times an hour for minute-long meals. Native to only North and South America (but with a range extending from Alaska down to Tierra del Fuego), the hummingbird is a little miracle. And we mean little: The bee hummingbird is the world's smallest bird; at two inches in height and weighing only .06 of an ounce, it is not much bigger than its namesake insect.

One who goes about life in a completely different manner is the majestic albatross of the southern oceans. The largest of the species, the wandering albatross, has a wingspan of up to 11 feet and weighs in at around 18 pounds; it maintains its strength on a diet of squid and fish. But unlike a hummingbird, an albatross exerts very little energy as it flies. An accomplished glider, it flaps its wings barely at all as it rides updrafts and thermals over the open sea. An albatross sails on the breeze for hours at a time; it can easily log 100 miles per day, and may cover well over a million miles in a lifetime. That life could last as long 60 to 80 years, which is 20 times the average for a hummingbird in the wild.

It is hard to believe that these two animals coexist as cousins in the avian fraternity.

FRANS LANTING/CORBIS

JIM BRANDENBURG/MINDEN

"Like a cross between a corkscrew and a jousting lance." That's how Arctic explorer Fred Bruemmer described the male narwhal's tusk in his 1993 book about the animal. The straight, spiraled ivory appendage that extends as much as nine feet from the upper jaw of this unique whale of the northern seas does indeed conjure notions of heraldic times. Specifically, it brings to mind the legendary unicorn—and that, centuries ago, got the shy but gregarious narwhal into a peck of trouble.

TUSK, TUSK

In the Middle Ages many cultures still believed in unicorns, which were seen as symbols of great purity and holiness—symbols, even, of Christ. The unicorn was ubiquitous in religious art of the period, and as Bruemmer writes: "All carry a horn that is unmistakably a narwhal tusk, the only long, spiraled horn in all creation." What had happened? Narwhal tusks were being harvested by unscrupulous hunters in the north, then were being trafficked as unicorn horns. These brought vast sums; the first Queen Elizabeth was given one that was said to be worth as much as a castle.

Gradually, of course, civilization came to terms with the fact that there are no unicorns, and this was a good thing for the narwhal (although jewelry makers continued to prize them, and still today Inuit hunters legally pursue narwhals). The global population of the animal is now between 25,000 and 45,000 individuals, and stable.

What is the purpose of the narwhal's unique tusk? That has long been a mystery, but a 2005 study by a group of scientists led by a clinical instructor from the Harvard School of Dental Medicine asserted that it is a dental extension with some 10 million nerve endings that acts as a sensory organ as the narwhal moves through the sea.

An Arctic neighbor of the narwhal who is equally renowned in the Brotherhood of Tusked Beasts is the walrus. A pinniped—a carnivorous, fin-footed, blubbery mammal like a seal or sea lion—the walrus is the only one to use its teeth to walk, pulling its 12-foot, ton-plus bulk across the ice with tusks that can grow to 30 inches. In the sea, where walruses spend two thirds of their time, the huge creatures become downright graceful, sometimes diving to depths of nearly a thousand feet at speeds of more than 20 miles per hour in search of clams and other mollusks. Walruses, like narwhals, are extremely sociable among their own kind, and to listen to male walruses singing their love songs during mating season—a cacophonous symphony that can be heard 10 miles away—is one of nature's oddest pleasures.

● The operative word seems to be swordplay, as a pod of male narwhals gathers at the Arctic ice edge in Canada's Lancaster Sound, where they dine on cod. Opposite: A Pacific walrus, also a male, poses for a portrait on Alaska's Round Island.

PAUL NICKLEN/GETTY

MATTHIAS BREITER/MINDEN

GETTY

He travels under different aliases (if he travels at all) in different locales. The oversized, hairy, apelike/hominid beast that haunts the northwestern United States and western Canada is known alternately as Sasquatch (from a Salish Indian term for "wild man" or "hairy man") or Bigfoot. To the Lakota Indians of northern Wisconsin he is Chiye-tanka, which means "Big Elder Brother." The creature that leaves prints in the snow of the Himalayas is the Abominable Snowman or, in the Tibetan dialect, the yeti. Other Asian regions know him as Meti, Shookpa, Migo or Kang-Mi. His counterpart to the north in Siberia is the Almas. As with the Loch Ness monster, his reputation relies on many reported sightings, some exceedingly strange photography and no hard scientific evidence.

ABOMINABLE BY ANY NAME

ROGER PATTERSON/CORBIS

Stories of the beast were already present in native traditions when British explorer David Thompson reported coming upon huge footprints in Alberta, Canada, in 1811. Similarly, B.H. Hodson, the first British Resident of Nepal, may have been informed by local legend when he saw, in 1832, something that "moved erectly, was covered in long, dark hair, and had no tail."

That description has held uniform in the decades since, as the size of the monster has settled into the 7-to-10-foot range. Some of the subsequent sightings were by people who, in other circumstances, would be judged serious-minded indeed. N.A. Tombazi of the Royal Geographical Society saw the Snowman during an expedition in the Himalayas in 1925; and in 1986, the famous Reinhold Messner, perhaps the greatest mountain climber of all time, saw the same. (He subsequently wrote that he thought the yeti could be the endangered Himalayan brown bear, which can walk upright as well as on all fours.) On top of the testimony came the photographs. One of an enormous footprint found in the Menlung Basin in the Himalayas in 1951 was particularly beguiling; it sold at auction for more than $7,000 in 2007, so someone still believes the elusive creature exists.

● Two famous "proof" photos: A print found by a 1951 British expedition in the Himalayas was seen as evidence of the yeti, while a picture taken in California in 1967, later understood to be a hoax, was supposedly of Bigfoot.

STRANGE BUT NOT TRUE

● Raphael's "Saint George and the Dragon," a small cabinet painting, was made between 1504 and 1506, and the famous tapestry "The Unicorn in Captivity" was woven, probably in Brussels or Liège in Belgium, at about the same time: between 1495 and 1505. Both of these European master-works are now in America. The unicorn hangs at the Cloisters in New York City, while the dragon is at the National Gallery in Washington, D.C.

O kay, we have to face the facts. (Well, we don't have to, but occasionally we should.) There are not now nor ever have there been any dragons. The Komodo dragon we met earlier is a monitor lizard that, although plenty fierce, neither flies nor breathes fire. Dragons are mythological. Trust us.

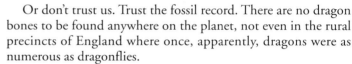

Or don't trust us. Trust the fossil record. There are no dragon bones to be found anywhere on the planet, not even in the rural precincts of England where once, apparently, dragons were as numerous as dragonflies.

You can wonder at Loch Ness (what's down there?) and the Himalayan yeti (what might be up there?), and still you have to disbelieve in dragons. Science says so.

Dragons are, along with unicorns, the superstars of mythical creatures. In *Beowulf,* written in the Middle Ages (between A.D. 700 and 1000), the anonymous author tells us: "The dragon began to belch out flames and burn bright homesteads; there was a hot glow that scared everyone, for the vile sky-winger would leave nothing alive in its wake." The template had been clearly established, so when Saint George battled his dragon and the Swiss knight Heinrich von Winkelried did the same, the two heroes were tussling with a beast whose physiognomy had been defined. The dragons of T.H. White, J.R.R. Tolkien, William Steig (the one in *Shrek,* by the way) and J.K. Rowling were little different.

The dragon was always, in Western tellings, representative and symbolic of evil, of Hades, of Satan himself. As we have already seen in our examination of the narwhal, the white-as-snow unicorn was the opposite, standing in for goodness, grace, Christ himself. In Eastern lore, interestingly, the dragon was often seen as a sacred beast, a unicorn with scales. These mythical creatures served myriad purposes.

To return to *Shrek* for but a moment: Ogres aren't real either. There have never been child-chomping giants of subhuman DNA. Charles Perrault referenced such beings in his 17th century *Tales of Mother Goose,* and the stories gained currency, but Perrault was trafficking in fiction.

As to brownies, dwarves, elves, goblins, leprechauns, pixies, fairies: fake, fake, fake, fake, fake, fake and fake. Perrault, Hans

MELLON COLLECTION

THE METROPOLITAN MUSEUM OF ART/ART RESOURCE, NY

Christian Andersen, Edmund Spenser and the ancients of Ireland made them fanciful and appealing, but they made them out of whole cloth.

And yet the cults developed—the believers in unicorns, in dragons, in fairies.

With the advent of photography—proof positive—they were able to boost their beliefs, and here we arrive at the dividing line between historic, classical myths and modern legends. Fairies—and perhaps Nessie and Bigfoot, depending on their reality or lack thereof—have a toe in both.

Beginning in 1917, in one of the most celebrated emanations of the otherworldly ever, Elsie Wright, 16, and her cousin Frances Griffiths, 10, cut out some fairy pictures from one of their children's books, fastened them to plants in the garden with hairpins and then posed amidst them. The so-called Cottingley fairy pictures, named for the English village where they were taken, caused an international sensation after photography experts and even Sir Arthur Conan Doyle, he of Sherlock Holmes fame, vouched for their authenticity. It was some 60 years before the cousins fessed up to their prank.

Many more hoaxes would follow. Westerners in the United States stuck antlers on a rabbit and had

ERIKO SUGITA/CORBIS

many postcard buyers believing in an animal called the jackalope. The legends of Loch Ness were exploited by conjoining two toys. Every April Fools' Day, you had to doubt whatever strange picture you might come across in a newspaper.

Or now, on the Internet. The most prominent and fun recent photographic hoax has involved Snowball the Monster Cat. In early 2000, Cordell Haughlie took pictures of his daughter's feline friend, Jumper, who weighed 20 pounds, then photo-manipulated one of them for about 20 minutes and sent the image of himself holding "Snowball" to some of his friends. The picture escaped into the wide, wide world of the Net, and soon was accompanied by a story of how the mother of the 87-pound Snowball had been raised near a nuclear power plant. It took Haughlie's (and Jumper's) appearance on nothing less than *The Tonight Show with Jay Leno* to set the record straight.

What will we believe in tomorrow? The sky is hardly the limit.

● Opposite: Two girls in the village of Cottingley borrow a camera and take a few photos, like this one, that happen to tap directly into English society's borderline-perverse fascination with fairies. None other than Arthur Conan Doyle boosted the pictures' credibility: "Matter as we have known it is not really the limit of our universe," he wrote. In 1983, the mischievous Elsie finally came clean about the fakes. But it seems astonishing today that anybody could have been buffaloed for so long a time. On this page, we have an ogre named Shrek, a giant mutant cat named Snowball and an anonymous jackalope, none of which are real but all of which are vastly entertaining—which is probably why we so want to believe.

EDWARD MCCAIN/AURORA

STRANGE DOINGS

CORBIS

SHELL R. ALPERT/U.S. COAST GUARD

In 1492, Christopher Columbus, sailing the ocean blue aboard the Santa Maria, saw an unidentified flying object "glimmering at a great distance." He wasn't the first ever to witness something strange in the sky and, Lord knows, he wouldn't be the last.

There were many more sightings worldwide in the 15th through 20th centuries, of course, but we must skip along here, so let's fast-forward to June of 1947, a seminal month in the annals of ufology. In one incident that occurred in the state of Washington, Harold Dahl was in his boat with his son and dog when he saw, over by Maury Island, six UFOs. According to Dahl, some hot slag fell from the spaceships onto his boat, killing the dog and injuring the boy. Furthermore, Dahl said, the next morning he was told by a so-called Man in Black, whom he took to be a military or government official, not to discuss the episode or some unspecified harm could befall his family.

ALIENS FLY AMONG US

hovered for 15 minutes near the Holloman Air Development Center in New Mexico on October 16, 1957, and was photographed by a government employee. Several such objects were captured in the morning sky over the Salem Air Station in Massachusetts by Coast Guardsman Shell Alpert on July 16, 1952.

On November 13, 1966, Ralph Ditter of Zanesville, Ohio, made one of the most famous-ever UFO photos with his Polaroid rendering of a classic-style flying saucer plaguing his house. He later admitted the picture was a hoax, done to please his daughter.

● Roswell: The incident and the industry. On July 8, 1947, Brigadier General Roger M. Ramey, commanding officer of the 8th Air Force, and Colonel Thomas J. Dubose, the chief of staff, displayed debris of what they said was a "weather device" (left). Their explanation cooled an overheated media (opposite, top) at the time, but has been questioned since. Today, at the International UFO Museum and Research Center in Roswell, kids can get up close and personal with alien models.

J. BOND JOHNSON/CORBIS

CHIP SIMONS

Only three days later, also in Washington, Kenneth Arnold was flying his small plane near Mount Rainier when he saw a squadron of nine mysterious conveyances traveling astonishingly fast and looking "like saucers skipping on water." When his account was reported the next day, flying saucers had a name.

The third incident, mysterious from the get-go, stands today as the most famous and controversial UFO episode ever.

On or about June 14, 1947, Mac Brazel, foreman of the Foster ranch, 70 miles north of Roswell, New Mexico, found some curious debris on the property. He eventually told Sheriff George Wilcox about it, and Wilcox informed Major Jesse Marcel at the Roswell Army Air Field. On July 8, the Air Field issued a press release stating that its 509th Bomb Group had recovered a wrecked "flying disc" from the ranch. Later in the day, official word came down that it was in fact a crashed weather balloon that had been found, and some of the debris shown at a press conference seemed to support this revised conclusion.

There the story might have ended had not Major Marcel granted an interview to ufologist Stanton T. Friedman in 1978, in which Marcel plainly stated that the military had covered up the recovery of a flying saucer. Ever since, the Roswell incident has been hotly debated, even as its narrative has continued to evolve and grow. Reports came forth that aliens had been discovered at the crash site, and that alien autopsies had been performed. (As might be imagined, an alien autopsy film surfaced, though it is widely regarded to be a hoax.) Serious folks got involved.

At one point, Barry Goldwater, senator from Arizona, hearing that UFO evidence was being kept in a top-secret place at the Wright-Patterson Air Force Base in Ohio, tried to gain access to that area and was denied. Goldwater, a former major general in the United States Army Air Corps, continued to press his request with General Curtis LeMay, who finally grew livid and, as Goldwater told *The New Yorker* in 1988, gave his friend "holy hell," telling him, "Not only can't you get into it but don't you ever

mention it to me again." Asked that same year by Larry King whether he thought the government was withholding UFO evidence, Goldwater replied succinctly, "Yes, I do."

Goldwater was by no means the only elected official to take UFOs seriously. During his 1976 presidential election campaign, Jimmy Carter recalled a 1969 sighting he had made in Leary, Georgia: "It was the darndest thing I've ever seen. It was big, it was very bright, it changed colors and it was about the size of the moon. . . . I'll never make fun of people who say they've seen unidentified objects in the sky."

There have been, in the decades since the incidents of '47, several official panels convened to look into UFOs, from which a number of reports have been issued. Project Sign was followed by Project Grudge and then Project Blue Book, which was headquartered at the Wright-Patterson base from 1952 to 1969. During this time, it investigated 12,000 sightings and determined about six percent of them could not be satisfactorily explained by known astronomical, atmospheric or human-caused conditions.

J. Allen Hynek, an astronomer at Northwestern University, was involved in all three of those official projects, and came to the conclusion that a very few of the soundest sightings did indeed imply extraterrestrial life. He was moved to found the Center for UFO Studies, whose "purpose is to promote serious scientific interest in UFOs and to serve as an archive for reports, documents and publications about the UFO phenomenon."

Of that there is no doubt: It is a phenomenon. It may be real, it may be fake, it may be something else entirely. But it is definitely a phenomenon, and has been for a long, long time.

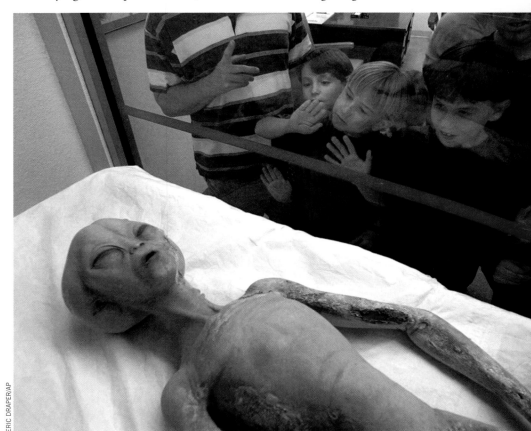

ERIC DRAPER/AP

Driftwood Library
of Lincoln City
801 S.W. Highway 101 #201

GIDEON MENDEL/CORBIS

A 1678 pamphlet was published depicting a woodcut that illustrated a story entitled "The Mowing-Devil: Or, Strange News out of Hartford-shire." It showed the devil mowing a round shape with a scythe in a farmer's field. The farmer insisted that this was what had happened: In the dark of night, some evil force had harvested his crop, and had done so in large, precise circular patterns. The farmer professed himself terrified, and who wouldn't have been?

CROPPING UP, BLOWING DOWN

Hartfordshire and Wiltshire in England have long been fertile breeding grounds for crop circles. What are they, precisely? They are enormous geometric formations made in grain fields by laying the growing plants flat. The big question is and has always been: What flattened all those stalks? These patterned shapes usually appear in the space of one night, and some observers of them insist that the manpower required to create them indicates that some unfamiliar force is at work. Extraterrestrials, with their whirring spaceships, have long been suspects, and the 2002 horror film *Signs,* starring Mel Gibson, popularized this theory.

All of Great Britain and much of the world was gaga over a rash of fantastic circles that appeared throughout the countryside in the 1970s, until two young men stepped up to take credit for several of them. Mysterious patterns still form on regular occasions throughout the world. They represent, collectively, an enigma in a hayfield.

Making crop circles seem like cute little pastoral curiosities is the Tunguska Event of 1908. At around 7:17 on the morning of June 30, something—a large asteroid? a comet fragment?—exploded approximately six miles above the earth's surface. Below, the region around Siberia's Tunguska River was devastated by the blast, which packed a hundred times the power of the Hiroshima bomb. Eighty million trees were blown down in an area of 830 square miles. It is fortunate indeed that the explosion occurred over a rural region. A blast of that size could have destroyed a city.

Witnesses to the event saw a bright light streaking across the sky only minutes before the explosion, but no one could say for certain what it was. Of course, there are some who hold that it was a spaceship.

The Mowing - Devil :
Or, Strange *NEWS* out of
Hartford - ſhire.

Being a True Relation of a Farmer, who Bargaining with a Poor *Mower,* about the Cutting down Three Half Acres of *Oats;* upon the *Mower's* asking too much, the *Farmer* ſwore, *That the Devil ſhould Mow it,* rather than *He.* And ſo it fell out, that that very Night, the Crop of *Oats* ſhew'd as if it had been all of a Flame; but next Morning appear'd ſo neatly Mow'd by the Devil, or ſome Infernal Spirit, that no Mortal Man was able to do the like.

Alſo, How the ſaid *Oats* ly now in the Field, and the Owner has not Power to fetch them away.

Licenſed, *Auguſt* 22th. 1678.

THE IMAGE WORKS (2)

● Crop circles are nothing new, as the woodcut (above) proves. But in our day, they have proliferated and become larger, more elegant and more complex, as illustrated by the snowflake design that appeared near an ancient burial site in Avebury, England, in 1997 (opposite). The incident that flattened much of Tunguska Forest (left) is unrelated to the crop circle phenomenon, though it produced a similar effect.

> "I know, also, that it has rained fishes in many places. . . . Certain persons have in many places seen it rain fishes, and the same thing often happens with tadpoles." So wrote the Greek Athenaeus in *The Deipnosophists* in the second century A.D. And so testified scores of other witnesses in the centuries that followed. And so scoffed the scientific community for the better part of the subsequent 1,800 years.

WONDERS FROM ABOVE

Then, in 1921, E.W. Gudger, an ichthyologist at the American Museum of Natural History, published in the respected journal *Natural History* a paper entitled "Rains of Fishes." Gudger's fellow researchers paid heed, and today it is generally accepted that small fish and amphibians have, in the distant and near past, become players in storms, transported on high by waterspouts or tornadoes. As Jerry Dennis wrote in his acclaimed 1992 book *It's Raining Frogs and Fishes:* "It is possible that a school of fish swimming near the surface could be sucked up, funneled high into the clouds by the rotating updraft, carried a few miles inland, and released with the rain as the power of the wind diminished. Theoretical calculations and measurements confirm that golf ball–sized hail requires an updraft of more than 100 miles per hour, which would be more than powerful enough to loft small fish high into a thundercloud."

Speaking of hail, it is only one of many other curiosities to be thrown down from the sky. Hailstones are generated by viciously powerful storms that propel water upward until it freezes, layer upon layer, and then, too heavy to climb further, the hail plummets. A hailstone can range up to four inches in diameter—the size of a baseball.

Lovelier than hailstones are snowflakes, yet another marvel of precipitation. Snow crystals are formed high above, then descend and join, collide and break apart, and coalesce anew. For all the trillions and trillions of snowflakes that have fallen throughout time, it probably is true that no two have ever been alike. To be so, as Dennis pointed out, "they would have to form in exactly the same conditions, collect the same number of molecules of water vapor in the same order, and bump into the same number of other crystals during their long descent to the ground." The snowflake is earth's ultimate individualist. And it is beautiful.

THE IMAGE WORKS

BLICKWINKEL/AURORA (3); PERENNOU NURIDSANY/PHOTO RESEARCHERS (1); RICHARD WALTERS/VISUALS UNLIMITED (1)

Opposite: A European woodcut from the 16th century depicts an unusual incident from 1355 in which frogs rained from the skies. On this page, several snowflakes seek to prove a point we learn in childhood: that no two are alike.

It is also known as the "Devil's Triangle," and that says it all. This is the region in the northwestern Atlantic where, supposedly, airplanes and ships seem to vanish. The boundaries are not formally drawn, but we can imagine the Triangle's apexes as San Juan, Puerto Rico; Miami, and, of course, Bermuda. The watery deep inside these points is haunted, or cursed, or simply a region that has been, over the decades, rife with ill fate.

BERMUDA'S TERRIBLE TRIANGLE

But is it fate? The area in question is extremely large (at least 500,000 square miles, using the most conservative measurement) and has long been heavily trafficked. It features the strong ocean currents of the Gulf Stream and is the breeding ground for violent hurricanes. In days of yore, pirates constituted a menacing presence in these waters. Furthermore, the Triangle was plied for centuries by sailors lacking radar and sonar, and is used today by boaters who buzz about in the waters off Florida and the Bahamas and throughout the Caribbean in pleasure craft too small to deal with a turbulent ocean. Is it any wonder that such a place has come to be thought of as dangerous?

Estimates as to how many tragedies have occurred in these waters vary widely; some say that since the time of Columbus's visits, there have been about 200 incidents, and others put the number five times higher. What can be determined by looking at the evidence is that the Triangle's notorious reputation rests largely on a few famous unexplained episodes from the 20th century.

In March of 1918, the USS *Cyclops,* built for naval use in World War I, was traveling from Barbados to Baltimore when it went down with 306 sailors aboard. On December 5, 1945, five Avenger torpedo bombers with 14 flyboys left Fort Lauderdale, Florida, on a practice mission over the Atlantic and were never seen again. In February of 1963, the SS *Marine Sulphur Queen,* with 39 crewmen and a cargo of molten sulphur, disappeared off the southern coast of Florida. All that was ever recovered from any of these incidents was a life preserver and some debris from the *Sulphur.*

Some researchers have cited an unexpected storm in the area in 1918; the probability that the men of Flight 19 became disoriented and ran out of gas in 1945; and a vessel that was unseaworthy in 1963.

Others have speculated on mysterious, perhaps paranormal forces at work.

In which camp are you?

CORBIS

● Right: The USS *Cyclops* met its end in the Triangle, as did the legendary Lost Squadron of Flight 19 (below). Opposite: In 1974, a posting on a Miami dock begs information about another victim: a 54-foot yacht carrying a crew of four.

NAVAL HISTORICAL CENTER

U.S. NAVY

JOE PAZEN/BLACK STAR

ALFRED EISENSTAEDT/LIFE

Integral to all religions is faith. The flock enjoys a system of belief: in their god, in their moral code, in what is being asked of them and what might be delivered. Sometimes, there is a belief in signs, symbols and miracles.

V I S I O N S

Christianity inspires, within the devout, an abiding faith. For many, this indeed extends to miraculous occurrences, and not only the ones involving Jesus Christ.

In the Pyrenean village of Lourdes in southwestern France, during a period extending from February 11 to July 16, 1858, the Blessed Virgin Mary is said to have appeared 18 times to 14-year-old Bernadette Soubirous. At one point during the ninth vision, the apparition revealed to Bernadette an underground spring and told her to drink from it. Not until her 17th visit did the lady reveal her identity. Mary's central message to Bernadette was, "Pray and do penance for the conversion of the world." The girl's experiences in the Massabielle Grotto were sanctioned as authentic by Pope Pius IX in 1862, and when the waters of the spring were declared to have miraculous properties, Lourdes became one of the world's principal centers of pilgrimage. At least two paralyzed men, Gabriel Gargam in 1899 and John Traynor in 1923, were said to have been completely healed by immersion in the holy water of Lourdes.

The Virgin is believed to have appeared in another European village—that of Fatima in Portugal—to three children in 1917. Today, Fatima is also a major site of pilgrimage.

As is Turin in Italy, where resides, in the royal chapel of the Cathedral of San Giovanni Battista, the Holy Shroud: a 14-foot, 3-inch-long piece of linen that seems to bear the bloodstains and body image of a crucified man of reasonable height. Was this cloth once draped over the crucified Christ? That question has been asked for more than six centuries, ever since the Shroud emerged in France in 1389. Upon being displayed publicly, it was denounced by the bishop of Troyes as a "cunningly painted" fake. The Roman Catholic Church has long been of two minds about the Shroud, accepting that it might be inauthentic when radio-carbon dating seemed to indicate the linen dates back only to the Middle Ages, yet respectful of what the image says about the suffering and sacrifice of Jesus. When the Shroud was last displayed to the public, for eight weeks in the spring of 1998, Pope John Paul II traveled from Rome to Turin to venerate the relic. The faithful took heart.

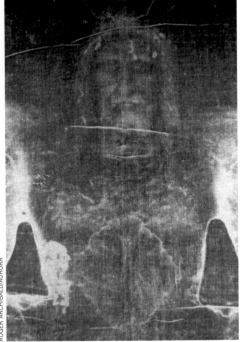

● Many pilgrims to Lourdes (like those, opposite, in 1958) come with hopes of being healed. The faithful who travel to Fatima (such as the ones above, in 1948) come to honor Mary's message. Those who trek to Turin on the rare occasions when the Shroud is on public display, do so to venerate—and wonder at—the image of a man. Is this the visage of Jesus Christ?

ROGER ARCHIBALD/AURORA

You might wonder whether the United States Department of Commerce doesn't have better things to do than authenticate haunted houses, but be that as it may: This agency of the U.S. government has indeed declared the Whaley House of San Diego to be officially haunted. This surprises few ghost hunters who know of its fearsome reputation; the Travel Channel's *America's Most Haunted* show has, in fact, anointed it the most be-spooked house in America.

The site's creepiness predates the stately Greek Revival house itself, built in 1856 and '57 by Thomas Whaley on (yikes! what a good idea!) ground where the city's public gallows had stood. Whaley and his wife, Anna, had three children, though (spoiler alert!) Thomas Jr. died in his upstairs bedroom at 17 months. A young girl died in this house, too, on the kitchen table after running into a clothesline with her neck.

So then: Thomas Whaley is often seen today on the second-floor landing, in his black frock cloak; when people stand between the parlor and music room, on the exact spot where Yankee Jim Robinson was hanged in 1852, they feel a choking sensation; from little Thomas's bedroom, cries issue; in the kitchen, a little blonde girl lives on, and pots and pans move of their own volition. Thus credits the Department of Commerce, an agency of these United States.

If the most haunted house in the U.S.A. is a once-private residence on the West Coast, its counterpart in the erstwhile evil empire, Russia, is a very famous public (sort of) building in the middle of Moscow. Lots of ghosts walk the halls of the Kremlin, most awful among them being that of the czar Ivan the Terrible, his face usually covered in flames. Seeing the apparition is said to have been an ill omen for his succeeding Russian leaders, which sounds reasonable.

Vladimir Lenin, too, has made several appearances. In fact, he appeared before he was dead: In October of 1923, he was critically ill in Gorky, but his spirit was seen digging through papers—perhaps tying up loose ends—in his Kremlin office. In 1961, Lenin's ghost contacted the medium Darya Lazurkina to let the Politburo know that he detested lying in state in the mausoleum right next to Stalin, which also sounds reasonable. Next day, Stalin was gone—interred elsewhere on the grounds. Now that's power!

SERGEI GUNEYEV/GETTY

KEVIN COOLEY

● The Whaley House (opposite) is not for the faint of heart. The Kremlin (above) has never been for the faint of heart, nerve or sinew, and some of its fearsomeness can be attributed to Lenin, founder of the Soviet Union, who today is safely (if not restfully) embalmed and entombed in the Red Square.

SERGEI KARPUKHIN/AP

Even the seagulls seem to want to keep a safe distance from Alcatraz. Opposite: In 1937, famed spiritualist Maggie Waite of Chicago, who had been coming to Lily Dale for four decades, achieves a trancelike state there. Waite's client list was said to include Theodore Roosevelt, Grover Cleveland, John Jacob Astor and the King of Belgium.

there are benevolent spirits and decidedly malevolent ones. No two places in America better represent this yin and yang of the spiritualism question than Lily Dale, a placid community on a lovely lake in Chautauqua County, New York, and Alcatraz, on an island in San Francisco Bay, now a museum in what was once a super-maximum-security penitentiary.

SERENE SPIRITS . . . AND LESS SO

The Lily Dale story starts with the sisters Kate, Margaret and Leah Fox, and in a New York hamlet named Hydesville. There, in the family home, the sisters claimed to hear mysterious rappings. They proceeded to set out patterns of knocks or finger snaps that were responded to in kind by the ghosts; the sisters were communicating with the dead. The three went on to organize demonstrations in public places (for which they charged, in the American way), and their fame spread. By 1855, they were seen as the founders of an American spiritualism movement, which quickly had a million adherents. In 1916, their childhood home was moved from Hydesville to Lily Dale, at the north end of Cassadaga Lake, and pilgrims began to flock. Today, the 500 residents of the charming, pretty town are all happy spiritualists, and travelers from the netherworld are equally at home. It is said that the wall between here and there is so thin in Lily Dale, even skeptical visitors are unnerved by . . . *something.*

If the spirits of Lily Dale are comfortable, those of Alcatraz never have been. In fact, the Bay Area's Miwok Indians refused to visit the island for fear of evil presences until 1859, when miscreant Miwoks were sentenced to spend time there in penance: the island's first prisoners. By 1912, a huge edifice, originally a fort but soon to become home to the country's most hardened criminals, had risen on the site. Al Capone, "Machine Gun" Kelly and "Doc" Barker checked in; few checked out in their lifetimes, and some maybe never. Capone can be heard playing the banjo, as he did in his day. In Cell Block C, an unearthly racket once grew so bad that the noted psychic Sylvia Brown was called in. She said a man named "Butcher" was to blame. Sure enough, when records were checked, Cell Block C was where mob hit man Abie "Butcher" Maldowitz had been murdered by a colleague inmate in the laundry room.

JON BRENNEIS/GETTY

REX HARDY/LIFE

It is often said—usually in Marfa, Texas—that the Marfa lights have been seen on a regular basis since the 1800s and that Native Americans of the region talked about them well before the first reported sighting by settlers in 1883. These ancient witnesses take care of, in the Marfans' view, any criticism that lights in the night sky might simply be the effect of distant headlights on Highway 67, or porch lights on an unseen ranch. They also ensure that Marfa, in rural west Texas, will enjoy some kind of tourist industry; that the Marfa Lights Festival on Labor Day weekend will be an ongoing affair; and that the Marfa Lights Viewing Area nine miles east of town will have a continued reason for being.

DO YOUR EYES DECEIVE YOU?

Go there, and look. From time to time, something weird appears and vanishes in the sky—basketball-sized, above the ground, bouncing like basketballs too. A dozen times or more each year, people say they have seen the lights after dark, just as folks in Hessdalen, Norway, see the Hessdalen Lights, and others in Queensland, Australia, see the Min Min Lights. An optical illusion, a trick of the local atmosphere? Or something otherworldly?

Folks elsewhere in the Southwest, in Sedona, Arizona, have long claimed that there is something strange in their atmosphere, too, though they haven't seen it. The so-called Sedona vortex is said to be a spiral of energy that acts upon a person in a way that promotes calming yin and energizes active yang. One totally groovy theory holds that Sedona's singular geology with its iron-rock formations rife with natural crystals, serves as a prism for the planet's electromagnetic energy. Whatever, ages ago Native Americans came to the area to pray, and today many people venture to Bell Rock, Cathedral Rock, Airport Mesa and Boynton Canyon to meditate.

There's something in the air over the desert, too, and in Arctic realms—and, in fact, over many hot asphalt roads in summer—that you think you see but you don't, not really. Mirages form when air masses of different temperatures near the ground cause incoming light to refract, or bend, causing reflected colors and shapes to be visible in an essentially empty horizon. You might see a boat in the sky, for instance. The Arctic explorer Robert Peary, coming over an ice ridge, swore he saw the mountains of somewhere he named Crocker Land stretching across the polar ice cap. But there is no such place, nor ever was.

THOMAS J. ABERCROMBIE/NATIONAL GEOGRAPHIC

RAUL TOUZON/NATIONAL GEOGRAPHIC

Opposite: A white cloud forms above Bell Rock in Sedona. The Marfa Lights (below) can be seen from an official viewing platform just east of town (right). Bottom: Heat waves create a mirage over the so-called Desert of Death in Afghanistan.

DANIEL SCHWEN

● Benjamin Spooner Briggs (right), captain of the *Mary Celeste* (below), and his wife and daughter were among those missing when the ship was found adrift at sea. His predecessor had already died aboard the ill-starred *Mary*. Opposite: As a woman, England's more fortunate Mary was immune to any bad vibes cast by the Kohinoor diamond, seen here in her crown.

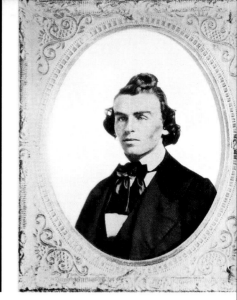

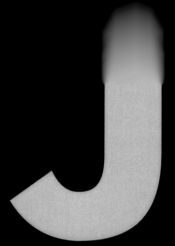

J

jinxes and curses are often difficult to credit, but sometimes bum luck so plagues an individual, object or place that one is forced to wonder. How many times must things go wrong aboard a ship before you start to blame the vessel itself? How much human tragedy must attend the possession of a coveted jewel before you're convinced that covetousness comes with a price? These are questions posed by the perplexing case histories of the brigantine *Mary Celeste* and the Kohinoor diamond.

CURSED

The 103-foot, 282-ton *Mary Celeste* (born as the *Amazon*) was launched in 1861 off Nova Scotia, and within days her captain died. Also during this maiden voyage, she collided with another ship in the English Channel; her hull was damaged, the other vessel sank. She then enjoyed a period of relative good fortune, but eventually ran aground back in her native waters (just by the way, the owner who tried to salvage her went bankrupt). She was repaired and returned to service in 1872.

That was the year of the (drumbeat) FATEFUL VOYAGE. On November 5, the *Mary Celeste* set out from New York City for Genoa, Italy, with a cargo of commercial alcohol. Days later she was sighted by another ship, drifting like a drunken sailor. The *Mary Celeste* was boarded and found to be deserted, save for her cargo and a store of food. Where had all the people gone? We likely will never know.

The 108.93-carat Kohinoor diamond, one of the world's most famous gems (and one that was even larger before being recut for brilliance on the orders of Queen Victoria), first came to historical light in 1306 when it was seized from the Rajah of Malwa in India. A Hindu legend that surfaced with the jewel said: "He who owns this diamond will own the world, but will also know all its misfortunes. Only God, or a woman, can wear it with impunity." That has proved to be about right.

The Shah of Persia possessed it on the day he was killed in a palace revolt, as did India's rulers on the day their land fell to the British. Indeed, all of the Kohinoor's owners have suffered ill fate or death, save three women: Queen Victoria, Queen Mary and Queen Elizabeth II, who has the diamond stashed today in the Tower of London.

THE IMAGE WORKS

"I bear in my body the marks of Lord Jesus." So stated Saint Paul, writing figuratively, in his *Letter to the Galatians.* Many centuries after Paul, in 1224, the charismatic Francis of Assisi, who himself would go on to be sainted by the Roman Catholic Church, was praying near the end of a 40-day fast on Mount Alverna in Italy's Apennine range when the marks of Jesus literally appeared on him. Brother Leo, a companion of his who was with him that day, later gave the first-ever clear testimony of stigmata being received: "Suddenly [Francis] saw a vision of a seraph, a six-winged angel on a cross. This angel gave him the gift of the five wounds of Christ."

S I G N S

CORBIS

● Above: Padre Pio in 1935. Left: On June 19, 2005, Romanian Orthodox nuns and monk Daniel Petru Corogeanu (at right) pray at the coffin of nun Maricica Irina Cornici, who died after being chained to a cross during a three-day exorcism. In 2007, Corogeanu was sentenced to 14 years for manslaughter.

Stigmata are manifestations—open wounds, feelings of pain—in the places on the body where Jesus Christ suffered the nails of crucifixion and, on that same fateful day, the soldier's lance. Some stigmatics have also experienced the pain associated with a crown of thorns. There have certainly been skeptics who viewed all stigmata as either delusions or self-inflicted frauds, but since Saint Francis's time, more than 300 people have been recognized by the church as having received "the gift." Among them are many saints, including Saint John of God and the mystic Saint Catherine of Siena. In modern times, the most famous stigmatic has been Padre Pio, who bore the marks as a young priest and went on to be the leader of a fervent following in his native Italy. In 2007, a book was published that suggested Pio may have faked his stigmata by using carbolic acid, but the Vatican countered that John Paul II had thoroughly vetted the late cleric, including all rumors surrounding the stigmata, before canonizing him Saint Pio of Pietrelcina in 2002.

The Catholic Church clearly believes that the devil can enter the body, and other religions have long held that Satan—or at least evil spirits—can enter and inhabit a person or place. To expel the malevolent presence, an exorcism, which routinely involves prayers as well as gestures and perhaps icons, is required. Jesus performed exorcisms, and today clerics in many faiths, ranging from Judaism to Hinduism to Islam to Scientology, do the same; in Christianity, John Paul II oversaw several exorcisms, and the Church of England still has official exorcists in all dioceses.

R

Rising from the dead is no mean feat, but there are those among us, human and animal, who seem to have turned the trick. In fact, Dosha, a 10-month-old pit bull mix from Clearlake, California, achieved it three times in one day in 2003.

You can look at the sequence of events as either terribly unfortunate or wonderfully lucky, depending on your point of view. First, Dosha jumped the four-foot fence that surrounded her yard and was hit by a pickup truck. "She wasn't moving and was glassy-eyed," neighbor Rolf Biegiela, who came upon the scene, told *People* magazine. "I said to myself, 'That's a dead dog.'"

A Clearlake cop said the same when he got there, and he shot Dosha in the head in case there was any misery to be put out of. The body was taken to the local animal-control center and put in the facility's freezer, pending disposal. Two hours later a worker opened the freezer door and there was Dosha, sitting up. She was promptly treated for hypothermia and had the bullet, which had just missed her brain and settled in the skin under her jaw, removed. Lingering effects: a bit of hearing loss, and presumably a good deal of trauma that, as a canine, she finds impossible to put into words.

Joan Murray suffered her own Doshaesque day, but if the misfortunes hadn't kept piling up, she might not be alive today. On September 25, 1999, the 47-year-old banker and recreational skydiver from Charlotte, North Carolina, exited a plane at 14,500 feet and, shortly thereafter, pulled her rip cord. Nothing happened. She had picked up speed to 120 miles per hour before she released her reserve chute, which opened and slowed her some, but then deflated. After 2.6 miles of plunging, she slammed into the ground at 80 miles per hour. Murray had landed squarely on a mound of stinging fire ants that quickly went to work; semiconscious and with the right side of her body wrecked, she suffered more than 200 stings before rescuers arrived. Her doctors said later that the stings may have shocked her heart sufficiently to keep it beating.

Murray lay in a coma for two weeks, but only six weeks after awaking, she limped home. Not long after, she was back in an airplane, heading up for another dive.

● Opposite: Back home in Clearlake after having survived her multi-chapter ordeal, Dosha shows some scars but is otherwise good to go. That phrase applies in spades to Joan Murray, who returned to the skies over North Carolina just as soon as she was physically able.

ACEY HARPER

DANNY TURNER

The world-record folks over at Guinness list hundreds of categories, but the ways in which a world record gets set basically come down to two: purposely or inadvertently. Fanatics worldwide jump through hoops to set a record, while others achieve one just by standing still and letting life come to them.

BUY THAT GUY A GUINNESS!

We could fill a Strange But True book with nothing but Guinness stories—as Guinness does on an annual basis—but we decided, instead, to highlight one of each type as an exemplar and a stand-in for all the rest.

Firewalking has been used in faith-based rituals since well before Christ, and as entertainment and competition in our time. How it works is complicated to explain, but suffice it to say, it is a good thing that coals burning at 1,000-plus degrees are covered in ash, a poor heat conductor, and that a bed of such coals provides an uneven surface, so that a fast-moving firewalker actually spends little time touching down. Even so, Scott Bell's fleet-of-foot feat is amazing. In January of 2006, the 36-year-old native of Cumbria in northwest England set his sights on Amanda Dennison's world record of 220 feet and shattered it, traveling 250. That effort landed him in the hospital for 10 days. "I've walked on fire thousands of times without any incident at all," he told the *BBC News.* "This is the first time I've had anything to write home about."

But not the last. Later that same year, in Wuxi City, China, Bell was back on the coals, tiptoeing a stunning 328 feet and thus extending his own record by just about a third.

Some might regard Scott Bell as a freak of nature. All of us can view the late Roy C. Sullivan in this light.

The so-called Human Lightning Rod, a career forest ranger from Virginia, was struck by bolts from the blue no fewer that seven times between 1942 and 1977. He lost a toenail in one incident, his eyebrows in another. He had his hair set on fire—twice. He was knocked unconscious. In the last incident, he was fishing when lightning sent him to the hospital with chest and stomach burns.

Sullivan died in 1983, the victim not of lightning but of his own hand. He committed suicide while in his 70s, this lucky, lucky man reportedly despondent at having been unlucky in love.

● Bell (opposite) is unfazed by fire; Sullivan (right) was unfelled by lightning (though his hat was the worse for wear). A note: These records, and that of the world's tallest man (page 14), are based on Guinness's 2008 volume. The mighty may have fallen since.

RANALD MACKECHNIE/GUINNESS WORLD RECORDS

AP

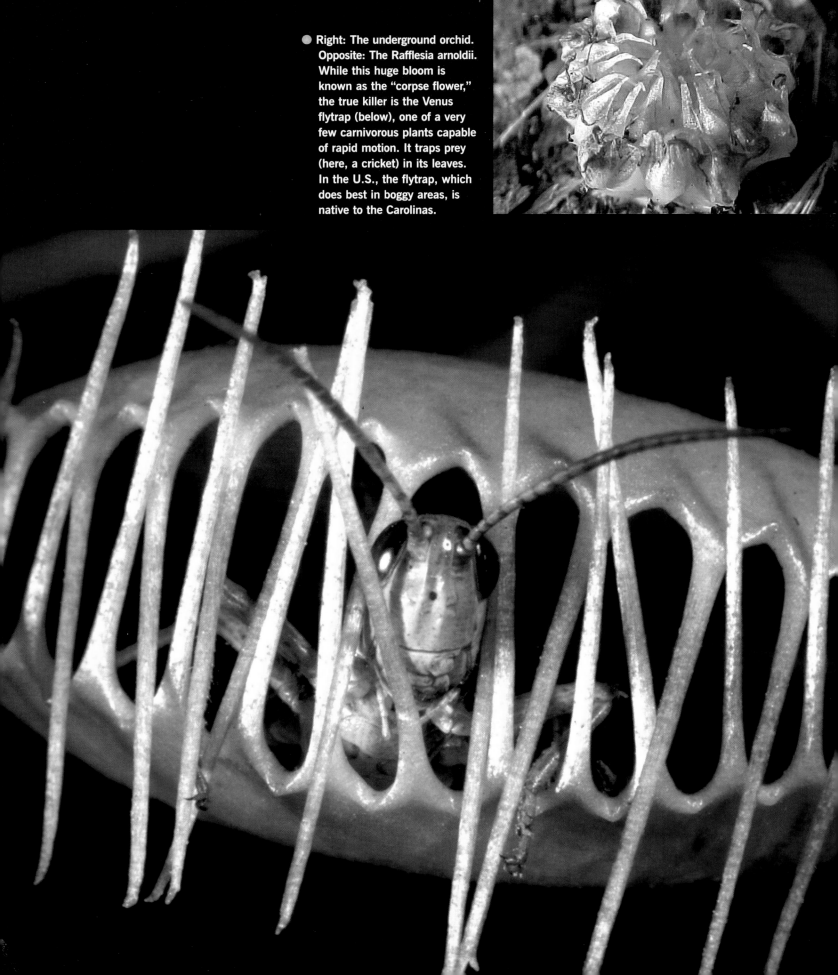

● Right: The underground orchid. Opposite: The Rafflesia arnoldii. While this huge bloom is known as the "corpse flower," the true killer is the Venus flytrap (below), one of a very few carnivorous plants capable of rapid motion. It traps prey (here, a cricket) in its leaves. In the U.S., the flytrap, which does best in boggy areas, is native to the Carolinas.

EVAN, ESTELLE, MALCOLM & ADELE CARRALL

There are weird plants everywhere, but we return to Australia, home of all manner of marsupials and odd mammalian animals, to visit a plant that is, it has been written, "so unique that it is considered a scientific wonder." This is the underground orchid, of which there are only two known species in the world; one is native to western Australia and the other is native to the east of the continent. To fully appreciate the flower's rarity even in its own family, consider that there are more than 35,000 species of orchids, the largest group of flowering plants on the planet; one of every seven flowering plants is an orchid.

VERY VEXING VEGETATION

The underground orchid was completely unknown in horticultural circles before 1928, when farmer Jack Trott saw a tiny crack in the ground and detected a sweet odor emanating from it. He removed a bit of soil and found a pretty white flower about a half-inch in diameter. This leafless plant, the western version of the orchid, had devised a way to grow and flower entirely underground (thus conserving water) in a harsh, desert environment. Between 1928 and 1959 the western orchid was found only six more times, in each instance purely by chance. Twenty years went by before it was seen again, and no new populations have been found since 1989.

The eastern underground orchid was first discovered in 1931 and has since been found in perhaps 10 places, most recently in 2002 by a 13-year-old boy. As it is still unknown what species of vegetation the orchid depends on for survival, the plants cannot be successfully transplanted or cultivated. So for now, this shy plant, which is certainly among the world's cleverest survivors, is still on its own.

The Rafflesia arnoldii, found in the rainforest of Indonesia, is at the other end of the spectrum from the tiny underground orchid. Also a rarity, its bright bloom is the world's largest, at three feet across and weighing as much as 24 pounds. You could put six or seven quarts of water into the hold in the center of the flower, not that you'd want to get that close. When in bloom, the Rafflesia arnoldii gives off an awful smell, like that of rotting flesh. (The odor attracts insects to pollinate the plant.) The scent is what has earned the Rafflesia its nickname: the "corpse flower."

DANIEL HEUCLIN/PETER ARNOLD

ALAIN COMPOST/PETER ARNOLD

M

Mythology sometimes overtakes truth even when dealing with such seemingly straightforward matters as those of state. Then again, sometimes matters of state become sufficiently surreal that the truth is stranger than any fiction.

THE HISTORICAL RECORD

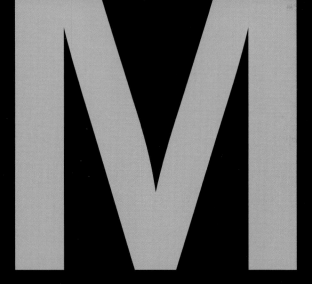

THANASSIS STAVRAKIS/AP

● Legends true and false: The Liberty Bell almost surely did not sound a clarion call to the people of Philadelphia in 1776, but did crack beyond fixing on Washington's Birthday in 1846. The Mask of Agamemnon (above) was probably not that of Agamemnon, though it did help confirm Troy's existence. The Dead Sea Scrolls expanded several Bible stories.

KENNETH GARRETT/NATIONAL GEOGRAPHIC

LARRY BURROWS/LIFE

Let's take a look at a few icons, both human and non, from the founding days of the republic. Consider, first, the Liberty Bell. It rang out brilliantly in Philadelphia on July 8, 1776, summoning the citizenry to Independence Hall to hear the first public reading of the Declaration of Independence.

Right?

Well, no, probably not. The hall's steeple was in such bad repair at the time that historians today doubt the bell rang at all during that seminal year. A fictional story in an 1847 edition of *The Saturday Currier* presented a romanticized version of the events of July 8, which gave the tale the patina of truth.

A far more accurate, eerier bit of lore attaches itself to the bell, however. Its famous crack was always in danger of growing too large to allow for any further ringing, and folks knew this. But at what instant did the crack say "nevermore"? This, from the February 26, 1846, edition of the *Philadelphia Public Ledger:* "The old Independence Bell rang its last clear note on Monday last in honor of the birthday of Washington and now hangs in the great city steeple irreparably cracked and dumb."

Was the Father of Our Country calling from beyond the grave? And were his compatriots in the cause of independence, fellow former presidents John Adams and Thomas Jefferson, calling to each other on, yes, Independence Day, 1826, when Adams succumbed in Quincy, Massachusetts (uttering his final words, "Thomas Jefferson survives"), even as Jefferson had died only a few hours earlier at his beloved Monticello in Virginia? It's true. You can look it up.

Sometimes, physical evidence turns yesterday's myth into today's historical record. Ancient Troy was not countenanced by most historians before Heinrich Schliemann's archaeological finds at Troy in 1873 and at Mycenae in 1876. And the Dead Sea Scrolls, discovered in 11 caves along the shore of the Dead Sea between 1947 and 1956, changed what we thought we knew about biblical times, expanding, for instance, the biographies of Enoch, Abraham, Noah and even God, among others.

What further clues or answers to life's great mysteries still lie buried out there?

Who can possibly know?

RAGNAR TH. SIGURÐSSON/ARCTIC IMAGES

They are otherworldly wonders, animated in the extreme but without animal spirit, a sensory experience beyond just the visual. They have been interpreted through the ages as omens, portents, signs and symbols. In fact, however, they are quite natural: an interaction of particles and gases that produce images of poetic beauty.

The northern lights are, in fact, as natural as air. Massive discharges of hydrogen from the sun transform into a gas of energized electrons and protons, and this "plasma" travels through space on the solar wind. Within five days of departing the sun's realm, some of the plasma reaches earth. Grabbed by the planet's magnetic field and pulled downward toward the north and south magnetic poles, the charged particles interact with atoms of oxygen and nitrogen, and juiced-up ions are produced. Energy is thrown off, and on the planet's nightside (yes, there are southern lights, too), the various wavelengths show themselves in arcs and draperies of green, white, yellow, red and turquoise light.

The science is interesting, the mythology more so. Canadian Inuit legend held that a heavenly light was emitted via holes through which the dead could pass. Alaskan natives considered the lights evil, while East Greenland Eskimos thought they were the spirits of children who had died at birth. Some medieval Europeans felt the glow reflected deceased warriors, while others read warnings of plague or conflict. Many cultures have seen the lights as some sort of game being played by the dead.

So, some things we know to be true; some we know are fanciful—and some we're still not sure about. Numerous witnesses swear they hear the lights hiss, whistle or crackle. Are the electromagnetic waves being transduced into acoustic ones? But the lights are occurring so high in the sky—how could the sound get to us so quickly, so in sync, albeit roughly, with the much faster traveling visual performance?

To be continued.

DO THE NORTHERN LIGHTS ROAR?

● Could any one of earth's creatures, including these lucky horses in Iceland, not be awed by the spectacular sight of the aurora borealis?

A playa is desert that occasionally becomes a lake. So sometimes it has a baked, hard-packed surface akin to smooth asphalt. Sometimes it has a slippery surface dusted with moisture. Sometimes it is a shallow body of water.

Racetrack Playa in California's Death Valley is very well-named. Once every two or three years, a rock of some size moves across the surface at an unknown speed—how fast *is* this racetrack?—and winds up in a place relatively distant. We know this because the stones leave long tracks that are just as telltale as the burned rubber of hot rods racing illegally down by the aqueduct. Because there are no adjacent prints, it is clear that neither humans nor animals are involved. But what sets the stones in motion and then keeps them going? No one has yet captured them sailing on film. Since the phenomenon was first noted in 1948, theories have included supernatural forces (of course), earthquakes and, in a now consensus view, wind.

With a slick but not submerged surface and a strong gust to boost the launch (plus sustained winds to keep it moving), a stone can travel a goodly ways—more than two football fields in length in the winter racing season (the best one for distance). Stones of 80 pounds and more have been known to make tracks.

While the sediments of Death Valley are conducive to sailing, the sands on the northwest coast of the Isle of Eigg in Scotland's Inner Hebrides are prone to sing.

This is a strange land before we get to its strange sand. In the Jurassic age, Eigg was a lagoon separate from the sea, where plesiosaurs (the prototype for the Loch Ness monster, according to some) gamboled. Then things dried out, leaving a classically Scottish, rough-hewn landscape and, in one place, a beach of pure, polished quartz crystals. (What caused this deposit remains a mystery.) In certain dry weather conditions, the sands squeak—they keen, they sing, they harmonize even. 'Tis a beautiful sound, and a haunting one.

SAILING STONES, SINGING SANDS

● In Death Valley (opposite) the rocks seem to come to life when heavy rains flood the playa and high winds supply the force. Right: The musical sands of Eigg are seen in the foreground. Those highlands across the Sea of the Hebrides belong to neighboring Rum Island.

MICHAEL MELFORD

PETE MARSHALL

Strange
but NOT
True

Was the nice Dutch colonial at 112 Ocean Avenue haunted? Well, none of the families who have followed the Lutzes to that abode since 1975 have had any problems with the paranormal. As for the "demonic boy" photo (opposite), it would have been possible to have staged such a shot. The most damning evidence that the whole thing was a hoax came from Ronald DeFeo's lawyer, William Weber, who said that after DeFeo's trial and conviction, he met with the Lutzes and discussed the possibility of concocting the haunted house story in hopes of getting a retrial for his client who, Weber would claim, had obviously been possessed by evil spirits.

The facts underlying some of the stories in this book, from the Loch Ness Monster to the event at Roswell, are admittedly debatable. But at least they are multisourced. When an individual or team fabricates a tale to prey upon the vulnerability of others, that is not good. Now, we're not saying this is what happened in Amityville, New York. We're not saying that the Lutz family exploited the beliefs and fears—indeed, the trust—of millions who might be persuaded to accept paranormal activity (though others have said exactly that). We're saying that if a story was concocted in hopes of reward in terms of books and movies, well, it worked beautifully. *The Amityville Horror,* based on the best-seller, was the second-highest-grossing flick of 1979, and no fewer than eight sequels followed.

CORBIS

George and Kathy Lutz moved into the house at 112 Ocean Avenue in Amityville in 1975 after having been informed that the notorious DeFeo murders—six family members dead, Ronald Defeo Jr. sentenced to 150 years in prison—had occurred under the prior ownership. The Lutzes lasted 28 days in the house, during which time, according to them, all manner of hell and haunting occurred. Then they moved out, and began to cash in. The evidence beyond their own testimony was scant, but George did put forth a photo of a presence with spectral eyes on the stairs. The photo was viewed by most as a fake, and this cast further suspicion on the whole story, as well as on the possibility that the Lutzes had, very early on, hatched a plan to profit from their "haunted" house.

GENE CAMPBELL (2)

C. WALKER/THE IMAGE WORKS

ERICH LESSING/ART RESOURCE, NY

● Jesus floats in the Korean clouds (top), and soldiers pose in one of the hokey "Thunderbird" photos floating on the Internet (above). In 1964, Priory of Sion founder Pierre Plantard appropriated the epigraph "Et in Arcadia ego," as inscribed on a sarcophagus in this circa 1640 painting (above, right) by Nicolas Poussin, as the Priory's motto, thus proving its ancient bona fides—or so Plantard hoped.

There have been other beyond-iffy photographs foisted upon the public that had nothing to do with money-making. Did Jesus really appear in the skies over Korea in the 1950s, seeming to bless a squadron of American B-29s in their war effort? No, probably not. The American pilot who shot the photo during a bombing raid in the north said that Jesus was in no way as distinct in the sky, and that he didn't know what he had taken before he saw the developed film. So was there doctoring done in the dark room? Perhaps. And besides, would Christ actually take sides in a death-making conflict?

Then there are the downright silly ones. What are we to make of the cinnamon bun in the Nashville coffee shop that looked like Mother Teresa, and became so famous that, in 2005, some nine years after it was "discovered," someone bothered to steal it? And what of the reflection of the Virgin Mary that was clearly visible in an office building's windows in Florida? Or the Jesus image baked, à la Teresa, into a piece of toast?

We are to make little of them, it is hoped, and yet many of us persist in accepting flimsy evidence or taking mere rumor as fact. There is a pervasive belief in various places around the country in a giant bird with a monstrous head and a wingspan of 20 feet or more that has been most often nicknamed "Thunderbird." It, like Nessie, certainly hearkens to the dinosaurs. There is reputed to be a famous and authentic Thunderbird photograph out there,

but unlike the many Loch Ness pictures, it seems to be as legendary as the bird itself—it cannot be found. It is not the one we display on the opposite page but a picture that supposedly ran in the *Tombstone Epitaph* in the late 1800s. So where is it?

Sometimes, witnesses—or readers—are convinced by "true believers" who are, to put it simply, wrong. We are not picking a fight here with the multitudes of *Da Vinci Code* fans when we say: The Priory of Sion is a hoax. It just is, no matter how much you might have enjoyed the book.

Pierre Plantard was a pretender to the French throne when, in 1961—the space race was already on, the Beatles were already a band—he boosted the Priory into some small sort of prominence. He gave it shape and definition. It was, theretofore, a French fraternal organization both founded and folded in 1956. Suddenly it was a vast secret society, a millennium old, involved in the reestablishment of the Merovingian dynasty in Europe—the legitimacy of which was based upon facts and claims that extended back to Christ. When the professedly nonfiction book *The Holy Blood and the Holy Grail* was published in 1982, the Priory gained adherents, and when *The Da Vinci Code* took off in 2003, it gained a fierce army of defenders.

But again: In 1961 it barely existed, and in 1955 it did not—not at all.

We're sorry about this, of course, but there you have it.

ERIK SIMONSEN/GETTY

STRANGE PLACES

THE SECRETS OF STONEHENGE

Stonehenge has remained the same for millennia, and yet it is ever-changing in our perspective. We are always learning something new about this ancient circle of standing stones, and what we learn serves either to confirm what we thought to be true or to quash the fantasies we harbored. We still do not know Stonehenge, not absolutely. Perhaps we never will.

MURDO MACLEOD/POLARIS

As much as any other entry in this volume, Stonehenge excites the sense of wonder in people. Everything about it is delightfully uncertain (well, almost everything, as we'll see). Where did it come from? Who created it? Why? When? And for heaven's sake, what is it? Armies of archaeologists and engineers, historians and ethnologists have tried to answer these questions, not always with complete success.

One aspect is without doubt: Stonehenge is no crop circle. It is certifiably old, and it was not built as a prank. It is ancient and it represents a very serious undertaking, one that required intelligence, purpose, strategy and persistence.

The word *Stonehenge* has an Anglo-Saxon meaning of "hanging stones." The composition began on the Salisbury Plain in southern England around 3100 B.C. when a ditch with a diameter of 320 feet was excavated, likely by Neolithic people using deer antlers. Evidence of such a primitive implement suggests the obstacles yet to be faced by Stonehenge's creators. Their ditch was banked, and along with other features there were two parallel entry stones. This construction, now often referred to as Stonehenge I, was in use for five centuries

before being abandoned.

Stonehenge II, commencing about 2100 B.C., saw the site dramatically altered. About fourscore bluestone pillars, as heavy as four tons apiece, were placed vertically in two concentric circles, which seem never to have been completed and were later dismantled. In 1923, it was established that the bluestone had likely been sourced from the Preseli Hills in southwest Wales, 240 miles distant. That would be quite a schlep today with a heavy rock; then, it must have been truly grinding. And yet it was clearly not dispiriting: Bluestone was the stone—the only stone—that would suffice for these people, so they went and got it.

About a century later, Stonehenge III presented a further

● Ancient and less so in Britain: Directly above and at top are the Avebury stones. Above, left, on Orkney Island in Scotland, is the Ring of Brodgar, which dates to circa 2500–2000 B.C. Opposite: The 131-foot-tall chalk horse of Cherhill may seem ancient but was cut in 1780.

transportation challenge. Although the quarry was a mere 20 miles away, the task was even more daunting as it involved moving sarsen stones as long as 30 feet and weighing up to 50 tons. Not content with simply building another ring, 97 feet across with a horseshoe shape inside, it was deemed necessary to top the circle with sarsen stones and then painstakingly pound the surfaces till they were wonderfully smooth. In subsequent years, several other devilishly complicated structures were assembled. The purpose of each of them remains open to innumerable interpretations, though one constant theory about Stonehenge has lately been confirmed: In all of its manifestations, it has been, as well as a place of ritualistic worship, a place of burial.

Whatever Stonehenge was originally devised for, it was still in use as recently as 1100 B.C. Old notions that it was built by Druids or Romans have been discarded; it was there long before those groups visited the neighborhood. Most theories envision Stonehenge as a place of worship of some sort, or perhaps as a kind of astronomical computer. The lay of the stones has always bolstered the latter theory, and the new evidence of burial grounds certainly supports the former.

The stones may have suffered the ravages of time and man, but people are more than ever magnetically drawn here, especially during the summer solstice when the sun rises perfectly over the heel stone. In fact, Stonehenge is such an attraction that the attendant hordes of visitors would doubtless have sent the devout ancients scrambling wildly for the hills—there, perhaps, to create another grand mystery among mysteries.

Which some of them certainly did even in their own time, with the Uffington White Horse chalk figure in Oxfordshire, which dates to the Bronze Age; with the earth mound at Silbury in Wiltshire, one of the world's tallest prehistoric man-made hills; with the stone circle at Swinside, in England's Lake District. And with the several more sites—at Knocknakilla, at Lisseyviggeen, at Newgrange, at Carrigagulla—in Ireland.

Stonehenge is eternal. It lives in and of itself, and in tributes paid from other formations ranging from those on the ground to those mounted by the band Spinal Tap. It is, perhaps, the world's most famous and beguiling mystery, renewing itself each morning with the sunrise, and not sleeping even as the sun sets.

DAVE BURGES/ZUMA

QT LUONG/TERRAGALLERIA

MARILYN BRIDGES

In the desolate—but nonetheless gorgeous—expanses of the western United States are cryptic formations, some man-made and some not, that shield mysteries. In desert realms in Colorado, New Mexico, Arizona and Utah, ancient dwellings are carved into cliffs, dwellings that once housed the Anasazi, ancestral Puebloans who were the Athenians of the Southwest. They lived in abodes up to 2,000 feet above the ground until the 13th century, only to vanish mysteriously.

IN THE WILD WEST

Who were they and how were they different from other Native Americans? We do not know with any certainty; we can't even know what these early Indians were really called. The term *Anasazi* means "enemy ancestors" in the Navajo language, and much of the scant record of the Anasazi comes from that tribe and other invaders, some of whom may have been implicated in the Anasazi's demise. It is also theorized that drought played a part. All that can be said is that they eventually either left of their own volition or were driven from their settlements, and by 1300 had all but disappeared. It was the job of ethnographers and anthropologists to determine that this once great nation became the Pueblo Indians, who still inhabit the Southwest.

And great they were indeed when they were the Anasazi. Petroglyphs, such as the evocative ones found on so-called Newspaper Rock in Arizona's Petrified Forest National Park, tell tales of daily life.

Clearly animals were important to these people as symbols as well as sources of food. They were cultured, productive and spiritual. In a paradox, the Anasazi were, in many ways, more sophisticated than their descendants.

Elsewhere in Arizona, there is a hole 570 feet deep and 4,000 feet across. This, too, like the Anasazi dwellings, hints at a tantalizing tale. In the 1920s, it was the first crater on earth confirmed to have been caused by an extraterrestrial impact. Subsequent evidence has filled out the story: Nearly 50,000 years ago, a massive meteorite as much as 150 feet across smashed into the planet at approximately 28,600 miles per hour. The sphere exploded with a force greater than 20 million tons of TNT.

In an odd postscript: The Standard Iron Company spent more than two decades searching in vain for the giant iron ball that was figured to be buried beneath the floor of Meteor Crater. We now know that the invading object was mostly obliterated.

JONATHAN BLAIR/CORBIS

GEORGE H. H. HUEY/CORBIS

● Opposite: Arizona's Meteor Crater is otherworldly after being blanketed by a rare snowfall. Above: It is thought that Chaco

This three-acre "great house" in Chaco, built between A.D. 850 and 1130, featured more than

F

● For the second time in this book, we present you with a hummingbird. But this one, as depicted in the Nazca drawing below, is 305 feet long. Opposite: The sun sets over Machu Picchu.

For centuries, secrets were harbored from the wider world on the plateaus and in the mountains of Peru. Then, early in the 20th century, spectacular discoveries were made, and mankind's sense of wonder knew no bounds.

CALLING TO THE HEAVENS FROM PERU

When airplanes began to fly over the land in the 1920s, pilots and passengers brought back reports of mammoth portraits of animals and strange geometric patterns etched into the canvas of the Pampa Colorado, or "colored plain," in the southern part of the country. Named after the nearest city, the Nazca Lines, which are virtually indecipherable at ground level, immediately fascinated and perplexed (and even scared) those who beheld them. There is a 360-foot-long monkey, a 443-foot condor, a 935-foot pelican, and there are nearly 70 more images of other animals, flowers, trees, plants and shapes: trapezoids, spirals, triangles and something that looks like a candelabra.

It was determined over time that the lines, which are to be found over a large region of nearly 200 square miles, had been drawn by people from the lost city of Cahuachi. The city arose nearly 2,000 years ago and was abandoned after 500 years. As can best be determined, the artists' methodology was to remove pebbles on the surface of the plain's desert to reveal the lighter color of the sediment beneath. The Pampa is among the driest and most wind-free plains on earth, and so the pictures were indelible.

But what do they mean? Theories have included: a giant astronomical calendar; a grouping of sacred paths; a vast outdoor temple; and, of course, airfields for extraterrestrial visitors (see the 1968 best-seller *Chariots of the Gods*). As a footnote: A portfolio of about 50 more giant drawings that is thought to have predated the nearby Nazca collection by some 500 years was discovered in 2005 near the Peruvian desert city of Palpa.

In 1911, a decade before the hidden-in-plain-sight masterworks of the Pampa Colorado were revealed, Machu Picchu, the Sacred Valley, the Lost City of the Incas, was found by Hiram Bingham in the Peruvian Andes. This 15th-century construction of 200 stone buildings on the plateau near the summit of a vertiginous mountain was a religious retreat of the highest order—highest, as in more than 7,500 feet above sea level. With its situation aloft, and with such iconography as the Intihuatana Stone, to which many visitors ascribe supernatural powers, Machu Picchu, too, has been seen as a place of outreach to alien life forms.

Or—more probably—to God.

ZUMA

GALEN ROWELL/MOUNTAIN LIGHT

STRANGE PLACES

O

One of the most remote inhabited islands on earth, Easter Island lies more than 2,000 miles from the shores of both Chile and Tahiti. It was settled in about A.D. 400 by Polynesians who, legend has it, were led by a chief named Hotu Matua, the Great Parent. When these early people arrived, their new home was lush, with giant palms useful for boats and housing. By the time their descendants were through, Easter Island would be a hellhole. Here, an entire culture was victim to an obsession.

THE EASTER PARADE

For some reason, beginning in the year 1000 and increasingly until 1600, the islanders became engaged in the construction and siting of statues, seemingly to the exclusion of anything else. They used volcanic tuff to carve at least 887 statues called *moai*, the largest of which was nearly 72 feet long. These *moai* were human forms, with exaggerated noses and ears. Because the islanders left no written record and but a flimsy oral history, there is no way to hold an informed opinion as to their purpose. In any case, simply carving these *moai* was not enough: A third of them were transported around the island and placed on *ahu*, ceremonial platforms that stood at an average of four feet high. Moving the *moai* was a tremendous undertaking—the tallest one erected was 32.6 feet high and weighed 82 tons.

Archaeologist Jo Anne Van Tilburg and others believe the islanders used palm trunks to move the statues. This deforestation became critical in the culture's decline and fall. With the disappearance of trees, topsoil was washed into the sea, leaving the inhabitants with no way to raise food; nor could they build boats to catch fish. Inevitably, the social order collapsed, and in its place came civil war and cannibalism. And in what must have been particularly frenzied activity, the *moai* were toppled by the Easter Islanders themselves. (Those standing today are the product of recent archaeological enterprise.) A passion in an isolated land thus led to obsession and madness, thence to extinction.

DAVID NUNUK/PHOTO RESEARCHERS

A horse greets the new day amidst half-buried *moai* statues on Easter Island (also known as Rapa Nui) in the South Pacific. As is the case in this picture, many of these marvels are on the slopes of the volcano from whose rock they were carved.

STRANGE PLACES

In the native tongue of the local Aboriginal people, the Anangu, who have officially owned it since 1985 (and who can be said to have unofficially owned it for millennia, even though land ownership isn't part of their creed), it is Uluru. It is also known as Ayers Rock, the name given to it by white Australian settlers after they came across it late in the 19th century.

It is perhaps the world's largest monolith, a bizarre rising of weathered sandstone in the hot interior of the island continent. It reaches a height of 1,100 feet above the desert floor, is 2.2 miles long by 1.5 miles wide, and is the above-ground extension of a larger underground rock formation in much the same way that the Hawaiian Islands constitute merely the summits of much larger mountains whose bases lie hidden far below sea level on the ocean floor. To the Anangu, Uluru is a sacred place that holds a record of the Dreamtime activities of the Ancestors; the tracks on its side are believed to have been left by those bygone natives. In a world filled with unique formations rising out of the ground, Uluru still stands out.

Below ground, our planet supports otherworldly realms. Familiar to many Americans are the stalagmite and stalactite cathedrals within Carlsbad Caverns in New Mexico and Mammoth Cave in Kentucky. Much less well known are Voronya Cave (also known as Krubera) in Abkhazia, Georgia's breakaway republic, and the Waitomo Caves of the North Island of New Zealand. Each is striking for a particular aspect: Voronya for its incredible depth (it's the world's deepest cave, at nearly 7,000 feet, much deeper, for instance, than the Grand Canyon), and the Waitomo system for its . . . glowworms.

Yes, glowworms. The Waitomo Caves, which may be as many as two million years old, have fine limestone incrustations, to be sure. But no other cave on earth has glowworms like these. Arachnocampa luminosa, which sounds like a spell cast by Harry Potter but is in fact a species of worm unique to New Zealand, hang from the ceilings of Glowworm Cave and Glowworm Grotto, casting a radiant, delicate filigree of light.

A deep dive into Voronya Cave is like a descent into Hades, while a slow cruise through Waitomo is a trip to Paradise.

NICOLE DUPLAIX/NATIONAL GEOGRAPHIC

● Uluru (left) resembles a loaf of bread baked by the sun of the central Australian desert. Here, at more than a mile down in Voronya, it seems like you're entering the gates of hell. Opposite: Waitomo's glowworms create a surreal heaven on earth.

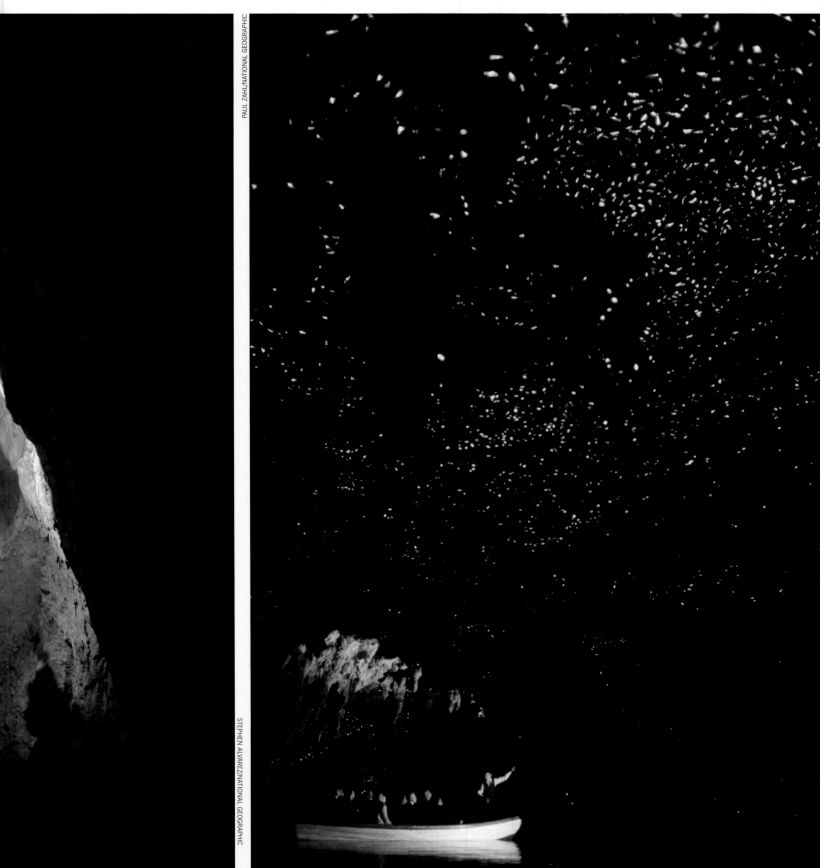

PAUL ZAHL/NATIONAL GEOGRAPHIC

STEPHEN ALVAREZ/NATIONAL GEOGRAPHIC

M

Morning Glory Pool is one of the most beautiful bodies of water in all of vast Yellowstone National Park—which covers 3,472 square miles in Idaho, Wyoming and Montana. The heated water in Yellowstone's basins allows vivid blue bacteria to grow year-round. The water is warm because beneath the terrain of Yellowstone is a 2,000-degree lake of molten rock, an underground pressure cooker 50 miles long and 30 wide. In other words, Yellowstone sits atop a gargantuan volcano.

WEIRD WATERS

For at least a few million years, since magma burned a hole under ancient Wyoming's bedrock, Yellowstone periodically disgorged a small ocean of lava in eruptions hundreds of times worse than any recent volcanic explosion on earth. Today, volcanic action heats rock below the surface, and this in turn creates all manner of thermal spectacle in thousands of steaming hot springs and hundreds of geysers (including the legendary Old Faithful, which erupts every 88 minutes), plus bubbling mud pots and fumaroles. Yellowstone, so extraordinarily peculiar, was never to be ignored: In 1870, it was set aside by Congress as the world's inaugural national park, the first place anywhere on the planet to be preserved simply because of what it naturally was.

No one needs to worry about preserving the Mariana Trench since its situation is so remote and forbidding that it can be visited only in sea labs customized to go to the greatest depths. The Mariana, near Japan, is one of at least 22 trenches in the world's oceans (18 in the Pacific, three in the Atlantic and one in the Indian). Actually a series of ocean depressions, it claims the title as the lowest of the low. It is in fact the lowest location on earth.

Just over 200 miles southwest of Guam is the Mariana's Challenger Deep, the very floor of the trench. Named after a British vessel that explored it in 1951, it bottoms out at 36,201 feet—seven miles down. Consider: If Mount Everest sat in the Challenger Deep, there would still be nearly a mile and a half of seawater above its summit.

Of course it is cold down there, but also warm in places due to conditions similar to those being produced in Yellowstone. Near hydrothermal vents, water is heated by acidic fluids from the planet's core to temperatures of more than 500 degrees. The odd situation supports unusual forms of life: The Mariana Trench is one of the domains of the bizarre anglerfish met earlier in this book (please see page 25).

⬤ **Left: Yellowstone's Morning Glory Pool is seemingly serene, though beneath its basin lies great turbulence. Far beneath the *Trieste*, a French bathyscaphe (in 1960, right), is the object of its search: the never serene Mariana Trench.**

BERNHARD EDMAIER

JOHN LAUNOIS/BLACK STAR

O

One is, supposedly, Treasure Island. Another is, purportedly, a country unto itself. A third was once a safe harbor for a legendary woman, or so her devotees want to believe. Where lies the truth?

ISLES OF INTRIGUE

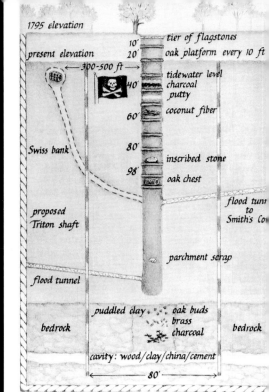

1795 elevation
present elevation

300-500 ft

10' — tier of flagstones
20' — oak platform every 10 ft

tidewater level
40' — charcoal
putty

Swiss bank

60' — coconut fiber

80' — inscribed stone

98' — oak chest

proposed
Triton shaft

flood tunn
to
Smith's Cov

flood tunnel

parchment scrap

puddled clay — oak buds
brass
bedrock — charcoal — bedrock

cavity: wood/clay/china/cement

80'

BIRUTA AKERBERGS HANSEN

PAUL NICKLEN/NATIONAL GEOGRAPHIC

The boys couldn't go further without help, so they returned with the Onslow Company and got down to 90 feet. At that point a stone inscribed with cryptograms was found. The marks will probably never be accurately translated, but according to one interpretation, they indicated that two million English pounds were buried another 40 feet below.

At the 90-foot mark, water began to seep in and by the next morning the pit was flooded. A booby trap in the form of a 500-foot waterway to the sea had been tripped. But also tripped had been the notion that there must be buried treasure here. Since the initial digs, companies backed by heavy hitters such as Franklin D. Roosevelt and John Wayne have searched Oak Island, to no avail. There's a cement vault at 153 feet, we now know, and an iron barrier at 171 feet—but still no treasure. Yet.

Redonda, a Caribbean islet, was sighted by Christopher Columbus on his second voyage, in 1493. The history of this steep and rocky spit of land might have ended there, except that it has been dubbed, from time to time, the Kingdom of Redonda—long ago by Caribbean eccentrics, and most recently by a pub in Southampton, England. It would help to have inhabitants, of course, which Redonda does not. It does have a Web site. And if that makes you a nation these days . . .

Across the world in the Pacific is Nikumaroro, formerly Gardner Island, barely four miles long and a mile wide. Did the aviator Amelia Earhart crash land there and survive on her attempted trans-world flight in 1937, or did she simply disappear into the waves? Will we ever know?

● **Opposite, top: A diagram of the booby-trap system at Oak Island shows the flood tunnels that have prevented diggers from reaching the bottom of the pit for more than two centuries. Bottom: Franklin Delano Roosevelt, third from right, helped fund a 1909 dig in the pit and remained fascinated by Oak Island. Above: the atoll Nikumaroro, some 1600 miles southwest of Hawaii, is theorized as Amelia Earhart's desert isle. Right: Redonda is home to no man or woman, but is claimed by off-islanders to be their personal kingdom.**

Oak Island is situated just off the Canadian province of Nova Scotia. On a summer's day in 1795, young Daniel McGinnis was wandering there when he came upon a depression in the ground. He had heard tales of pirates on Oak Island, and of the possibility of buried treasure, so he went and got a couple of his fellow teens and they started digging.

They were thrilled when, two feet down, they found flagstones covering a circlular hole. At 10 feet, oak logs. At 20 and then again at 30 feet, more logs.

FREDERIK RAMM

At seven o'clock on a March morning, the azaleas in one of the old sections of Bonaventure Cemetery make for lovely accents against the moist drip of the Spanish moss and the early gray fog. Meanwhile in the dog chapel in Vermont (opposite), canine accents are everywhere, including in the stained glass.

To many people, a graveyard offers solace: a chance to visit—perhaps commune with—a lost family member or friend. But for many others, a graveyard represents an eerie experience without rival. In a cemetery, the veil between life and death is simply too thin. There, if our world indeed supports ghosts, is where the dead sleep. Or refuse to.

A PRAYER FOR THE DEPARTED

Literary lions from Edgar Allan Poe to Stephen King have long appreciated the value of a graveyard setting in establishing a creepy mood. When the writer John Berendt looked into the strange events surrounding a death in Savannah, Georgia, in his celebrated 1994 non-fiction book, *Midnight in the Garden of Good and Evil,* he recounted nocturnal visits made to (and voodoo rituals performed in) Bonaventure Cemetery. In point of fact, Bonaventure is an excellent example of a haunted graveyard. The moss-draped oaks inside the wrought-iron gate look like close kin to the animated trees of *The Wizard of Oz* or *The Lord of the Rings.* The statuary is extraordinary, even by stringent haunted-graveyard standards. Angels, praying children and grieving women are among the stone denizens of Bonaventure.

Plentiful, too, are the stories. Josiah Tattnall Jr. once owned a plantation on the site of the present-day cemetery, and during a dinner party, the house caught fire. The owner ordered his slaves to resituate the festivities outside, and today Tattnall's guests can sometimes be heard chatting cheerily and toasting with fine crystal. More unsettling is the growling of Bonaventure's resident pack of angry ghost dogs. One of the cemetery's best-known monuments, the statue of Little Gracie Watson, has been said to cry out in the night. Another, that of the so-called "Bird Girl" (who died of pneumonia when she was 6), became so popular an attraction after serving as the cover image for Berendt's best-seller that it was moved indoors to the Telfair Museum. A graveyard, after all, should be a place of peace and quiet.

The tiny Dog Chapel in St. Johnsbury, Vermont, is just that—in fact, the canines here are considerably less demonstrative than those in Bonaventure. Built by the artist Stephen Huneck after he survived a life-threatening illness in 1998, the chapel welcomes "All Creeds, All Breeds—No Dogmas Allowed." It is inarguably the perfect place to mourn a late pet, and also to celebrate the value of man's best friend.

LORI SMALTZ

PAUL O. BOISVERT/NEW YORK TIMES/REDUX

The Great Pyramid at Giza is the only one of the Seven Wonders of the Ancient World still standing. In the eyes of modern man, it is the original Wonder, the Wonder from which all others descend. Prehistoric tribes on unknown islands surely constructed remarkable cities and spectacular totems to their god or gods. But this, a monumental tomb that Egypt's King Khufu ordered more than 4,500 years ago to honor and accommodate none but himself, is *the* Wonder of Wonders.

FIT FOR A KING

Khufu (circa 2551 to 2528 B.C.; also known as Cheops or Suphis) ruled for two dozen years during the Fourth Dynasty (2574 to 2465 B.C.) and was rumored to have been quite a tyrant. Beyond that, and the fact that he was the son of a pyramid-building king named Snefru, precious little is known. He clearly approved of Egypt's Old Kingdom tradition of erecting pyramid-shaped superstructures to encase royal tombs, and chose the Giza Plateau, on the west bank of the Nile across from Cairo, as the site of his own.

It is believed that 100,000 men, probably farmers displaced by the river's annual floods, worked on the pyramid for up to four months each year for at least 20 and perhaps 30 years. The word "toil" doesn't begin to tell the tale. Consider that the pyramid's 2,300,000 blocks of stone average 2.5 tons apiece (the heaviest are 15 tons; the smaller ones are placed nearer the top);

consider that the blocks needed to be dragged by teams of men up ramps made of mud and brick, and that the ramps then had to be raised for each new level; consider that only the interior stones were quarried nearby because Khufu's architect wanted a better quality white limestone for his casing layer, and that this high-grade rock needed to be brought from quarries on the other side of the Nile at Tura, then up from a harbor at the plateau's edge; consider that dense granite blocks were needed for the burial chamber in the pyramid's center and to serve as plugs for the corridors, and that the heavy granite could be supplied only by a quarry in Aswan, nearly 600 miles south; consider that the pyramid as designed was required to grow to a height of 481 feet and that its base spread over some 13 acres . . . and then consider that ancient man, through blood, sweat and a considerable flow of tears, pulled it off.

PETER MILLER/GETTY

Khufu's Great Pyramid does not lack for structural company on the plains surrounding the mighty Nile. There are many other pyramids, plenty of sphinxes and, in the Valley of the Kings across the river from ancient Thebes (today's Luxor), the tombs of dozens of royals and nobles from Egypt's New Kingdom.

BILL ELLZEY/GETTY

● The Great Sphinx gazes upon the Giza Plateau from its place near Khafre's pyramid; many believe the sculpture bears Khafre's features. Below is a section of the antechamber of Tut's tomb, and opposite is the triumphant moment as Howard Carter (squatting) opens the doors of the fourth shrine. The winged tutelary gods carved on the doors indicate that Tut's pursuers are reaching the heart of the matter, wherein lies the sarcophagus.

I N S C R U T A B L E

The Greek word *sphinx* means "strangler," and it applied originally to a hybrid with the head of a woman, the body of a lion and the wings of a bird. The so-called Great Sphinx, with its man's head, lies near Khufu's pyramid, hard by that of his son Khafre. For centuries the theory was that this 240-foot-long, 66-foot-tall statue was a second tribute to the lad. But in the late 20th century, doubts were raised about the monument's age. The Great Sphinx has its secrets, and isn't talking—not about this, nor about the rumor that there is evidence of a lost (and advanced) civilization hidden under its front right paw.

The expansive necropolis in the Valley of the Kings was constructed over approximately 500 years, from the 16th to the 11th centuries B.C. The British archaeologist Howard Carter had already supervised excavations that revealed the tombs of such as Hatshepsut and Thutmose IV when, in 1922, working in collaboration with the amateur Egyptologist George Herbert, fifth Earl of Carnarvon, he first pried open a sealed doorway, and, with candle in hand, gazed upon the treasures of Tutankhamen. "It was sometime before one could see, the hot air escaping caused the candle to flicker," Carter wrote in his diary, "but as soon as one's eyes became accustomed to the glimmer of light the interior of the chamber gradually loomed before one, with its strange and wonderful medley of extraordinary and beautiful objects heaped upon one another." Over the next 10 years, Tut's tomb, one of the most exciting finds in history, was excavated.

In 1699, the first horror story involving a mummy's curse was published, and the idea attached itself through the centuries to many entombed pharaohs and kings—but to none more strongly than to Tut. It is true that Lord Carnarvon, principal sponsor of the dig, who had entered the tomb at the side of his man Carter, was bitten by a mosquito in the weeks following, and became ill and died at age 56. Tut, the boy king, remains innocent until proven guilty.

HARRY BURTON/CORBIS

GETTY

E Easter Island is a grand collection of enormous sculptures that testify to a magnificent communal obsession. There exists 30 miles south of Miami a more modest but no less confusing product of a deep, deep passion, this one belonging to a single man. The Coral Castle is Edward Leedskalnin's paean to love unrequited, and it is his life's legacy.

INDEPENDENT MINDED

KIM GILMOUR

ANDRE JENNY/THE IMAGE WORKS

● When Sealand's Roughs Tower (above) went into service during World War II, not even the most imaginative seer could have predicted that half a century later, it would be advertising not only its sovereignty on its flank but also its Internet address. When the Coral Castle was begun in 1920, it was a most private, personal enterprise. But such a wonder was clearly predestined to one day entertain tourists.

In 1920, after the Latvian immigrant Leedskalnin had been left standing at the altar by his 16-year-old fiancée, he did two things with emphasis: He set to work and he became a recluse. Because of the latter, we are left to wonder at details of the former. What we do know is: Until his death in 1951, Leedskalnin reportedly quarried and put in place more than 1,000 tons of coral rock and fashioned his castle and its fabulous accoutrements. We say "reportedly" because the spurned suitor worked mostly by night, and engineers who have visited the castle subsequently are hard pressed to see it as the work of one man, no matter the intensity of his passion.

Today, in the great postmodern pop culture tradition, the castle has been commemorated in song—rocker Billy Idol's "Sweet Sixteen"—and turned into a tourist attraction.

Sealand, by contrast, cannot accommodate tourists, but is attractive in quite another way.

Its story begins during World War II: To defend against German air raids, England erected a number of sea forts along its eastern coast, each capable of supporting a limited number of men and their artillery. Just north of the Thames River estuary was the concrete-and-steel Roughs Tower. International legalities were more likely to be disregarded in those days of duress, and Roughs Tower was built seven nautical miles from land, more than twice the three-mile limit that was England's territorial privilege. In other words, the fort was in international waters.

In 1967, Roughs Tower, having been abandoned by the military, was occupied by a former British Army major, Paddy Roy Bates, who enlisted some English barristers to back up his claim to a sovereign state. He declared Sealand a principality, and himself and his wife the prince and princess thereof; the remainder of his family constituted the royals. He has had to fight invaders from his homeland and from Germany and the Netherlands—not to mention the courts—ever since. But Sealand lives, a sort of super-Redonda with actual citizens.

This tale is often entitled The Legend of Atlantis, but that is doing the famous lost continent a disservice. Yes, it has never been irrefutably found, it has not been "discovered" à la ancient Troy. But its story was first rendered by one of the world's greatest-ever seekers of—and purveyors of—truth. If Atlantis is an invention, Plato was its inventor. If it truly existed, Plato was its historian.

IN SEARCH OF LOST ATLANTIS

HERBERT LIST/MAGNUM PHOTOS

GETTY

In 360 B.C., the Greek philosopher wrote of an idyllic island "greater in extent than Libya and Asia" and of an advanced civilization that had become corrupted and was finally destroyed by an earthquake or tsunami. Atlantis was situated vaguely by Plato. Most analysts have posited that when he wrote of "those who dwelt outside the Pillars of Hercules," he meant a people who inhabited a land that existed somewhere in the Atlantic. And Atlantis has, over the years, been "found" in or around Cuba; on Antarctica; in the north Atlantic near Iceland. Famous believers have included everyone from the psychic Edgar Cayce (who listed toward Bimini) to biologist Rachel Carson (who favored the north).

Few of the nominated places possess something close to what Plato described as Atlantis's fate, in their historical records. But there is a notable exception, and while it wasn't nearly as large as "Libya and Asia," it was in Plato's neighborhood.

We can only guess what precisely happened on the Mediterranean island of Thera since nothing about the event, which took place circa 1470 B.C., was written contemporaneously. After examining the evidence, though, archaeologists are certain that the ring of five islands now known as Santorini was once part of a single island that was 10 miles wide and dominated by a mountain nearly a mile high. The top of that mountain exploded in titanic fashion nearly 3,500 years ago, covering the island in ash and sending a tsunami as high as 300 feet crashing against the cliffs of Crete 70 miles to the south. The sonic boom emitted could be heard deep in the African jungle and along the Scandinavian shore. Thirty-two cubic miles of Thera were obliterated, and damage everywhere throughout the Mediterranean was surely tremendous.

Some have speculated that the immensity of the calamity might have contributed to the swift decline of Crete's sophisticated Minoan civilization. Others have theorized that stories of what occurred on Thera, handed down by word of mouth, found their way to Plato.

And perhaps they did.

ERICH LESSING/ART RESOURCE, NY

● Opposite: Nea Kaimeni and Palaa Kaimeni, as seen from Santorini, are all that are left of Thera's vent. Left: Did this wall painting from the 13th century B.C., found among the ruins of Thera, depict daily life in an Atlantean port? Above: The Sacred Way on Santorini, too, dates to the era when the island was known as Thera.

M

Municipalities around the globe participate in sister cities programs, in which the people of one community pledge fealty, friendship and the promise of a welcoming handshake to those of the other. Boston, for instance, is a sister city to not only sublime Kyoto and sexy Barcelona but to steady-as-she-goes Melbourne. No cross-boundary relationship is stranger, though, than the one that has been forged without any official acknowledgment between Beijing and a south Texas city called Katy.

● Below: A view of the Wumen, or Meridien Gate, the entrance to the inner sanctum of the Forbidden City. As for the Terracotta Army: The one in China dates from 210 B.C.; it was discovered in 1974 and has an estimated 8,000 soldiers. This one in Texas is a mite smaller and a lot newer but is still pretty impressive.

Beijing, you might know, is home to the Forbidden City. Smack dab in the middle of China's capital is a collection of nearly a thousand buildings that are some 600 years old, the largest surviving palace complex and largest assemblage of ancient wooden edifices in the world. This is the ominously named Forbidden City, the imperial palace and court during the Ming and Qing dynasties. Construction of the city, which is set in a rectangle 180 acres in size, began in 1406 and (as can be accomplished in a dynasty) was completed in a remarkably short time, 14 years. A million workers, including 100,000 artisans, were forced into hard labor in Beijing; forests in faraway provinces were felled for their timber, and tremendous stones quarried outside the city were slid into place over the winter ice. Through the centuries, the Forbidden City was attacked, looted and ransacked several times, but it served as a palace right up until 1912, when Puyi, the last emperor of China, abdicated. The city today, well preserved and housing a fascinating museum, is a glorious if haunting place.

The Forbidden Gardens outside Katy are, by contrast, a grandiose and entertaining rendering of the Chinese story. They were built in 1997 on some 40 acres for a lot of loot (as can be accomplished in a capitalist society). Ira P.H. Poon, a Hong Kong real estate mogul, spilled an estimated $40 million across the Texas flatlands to realize his dream: a vivid telling of more than 2,000 years of Chinese history. Emperor Qin's famous terracotta soldiers, 6,000 of them, are replicated at one-third scale; the Forbidden City itself is a twentieth the size of the real deal. There are shaded courtyards and koi ponds providing relief from the blazing summer sun, as Chinese zither music floats forth from hidden speakers. Here is exotic China, Texas style.

FORBIDDEN FRIENDS

DENNIS COX/CHINASTOCK

DAVID WOO

The border of northern Tibet is defined by a long range of mountains, the Kunluns, that feature peaks of up to 23,000 feet and hidden-away, idyllic valleys. This world apart has long held a special place in Taoist lore; a Han Dynasty religion led by the Spirit Mother of the West was once headquartered in the Kunluns, or so it's said. There is, it has long been believed in the region, magic in these mountains.

Now, let us add to this some interesting facts. The people of this corner of Asia are not only relatively long-lived but extremely vital and illness-free in their old age. Eating vegetables, usually raw, on a daily basis and exercising naturally in the rhythm of an agrarian life, the natives of this part of the world are not only healthy but by all accounts generally happy.

And so, of course, the Kunlun Mountains have been regularly put forth as a candidate for Shambhala, the pure land of Tibetan Buddhist tradition whose name translates from the Sanskrit as "place of peace." For the popular worldwide audience, British author James Hilton reinterpreted Shambhala as "Shangri-La" in his 1933 novel, *Lost Horizon*. This isolated place where people lived in contentment and seemingly forever was, instantly, an intoxicating idea. And in the Kunluns, the seeds of that idea live on today.

In the ancient Cambodian city of Angkor, more than 100 stone temples, built between A.D. 802 and 1220, also live on—as not only astonishing relics but as still-operative houses of the holy. There are said to be refuges for the spirits of the uncomfortably dead as well. The countless thousands massacred in modern times by the vicious guerilla movement, the Khmer Rouge, particularly haunt Angkor.

Thick jungles surround the area around the temples, and for centuries this place was as remote from the Western World as were the anonymous plateau sanctuaries of the Kunluns. But since its discovery by French missionaries in 1860, Angkor has stood as one of the great, mystical wonders of our planet.

W.E. GARRET/NATIONAL GEOGRAPHIC

LANDOV

Opposite: High in the Kunluns, large glaciers dominate the Qinghai-Tibet Plateau. Below: Two Buddhist monks in a doorway of the Bayon Temple. The 12th-century edifice, with more than 200 large faces adorning 54 towers, was one of the last temples built in Angkor.

Greek mythology is so rife, from top to bottom, with fantastic tales—of gods battling one another in the sky, of Zeus hurling his thunderbolts earthward, et cetera—that it is just as hard for modern man to believe in the settings of those stories as it is for him to credit the plotline. But Mount Olympus, where dwelled many of the gods and goddesses worshiped by the ancient Greeks, is a very real mountain.

And it is an impressive one at that. From its base at sea level in northeastern Greece, it rises to a peak of 9,570 feet on the highest summit, Mitikas, making Olympus not only the country's highest mountain but one of Europe's most impressive in terms of pure vertical ascent. Way back when, the mountaintop was home to the Twelve Olympians, the principal gods in the Greek pantheon. Zeus and his wife, Hera, were in residence, as were Zeus's brothers, Poseidon, god of the sea, and Hades, ruler of the underworld, as well as his sister Hestia, goddess of the hearth. The gods' great houses near Olympus's summit had been built by Hephaestus, god of the forge, and entrance to the community was through a gate of clouds overseen by goddesses called the Seasons.

Down below, the ancient Olympic Games were staged every four years in Zeus's honor. His primacy among the gods in the eyes of the Greeks was confirmed in 2003 when a pre-Christian-era temple dedicated to him was discovered at the foot of the mountain, with an insignia referring to Zeus as "the highest." This indicated that the Greeks were moving from polytheistic religious beliefs toward a monotheistic philosophy—a thinking that would be espoused by Christianity as it rose and spread throughout Europe.

Much of the thunder associated with Olympus must be regarded as figurative. Not so the thunder of Ol Doinyo Lengai, an active volcano in the African Rift Valley in Tanzania that is also home to a god, in this case the Masai deity Eng'ai. Called "the strangest volcano on earth" by *National Geographic* magazine, it spews unusually cool (1,000 degrees Fahrenheit) and fluid lava that can harden within seconds of being extruded. The mountain is therefore decorated with natural formations that seem to have been shaped by the legendary Catalan architect Gaudi or perhaps by, as *National Geo* had it, Dr. Seuss.

MARK DAFFEY/GETTY

● Opposite: Trekkers can be
seen just to the right of the
ridgeline as they make their
way toward the Mitikas
summit on Mount Olympus.
These clouds once
represented the portal to
the community of the gods.
Above: A time-lapse
photograph shows lava
flowing on Ol Doinyo Lengai.
The liquid is destined to
spend just a short time in
the open air before solidifying
into natural sculptures such
as this whimsical wing (right).

CARSTEN PETER/NATIONAL GEOGRAPHIC (2)

Tor is a Celtic word meaning "conical hill," and a perfect example is Glastonbury Tor where, perhaps, Arthur once roamed; evidence of a 5th-century fort has been found there. Opposite: The hollow Shambles Oak in Sherwood Forest was also called Robin Hood's Larder, as legend held that the outlaw stored his venison in it. The great old tree blew down in 1962.

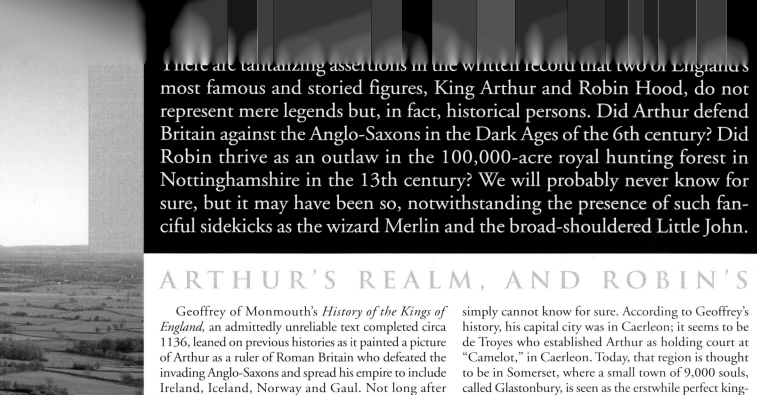

There are tantalizing assertions in the written record that two of England's most famous and storied figures, King Arthur and Robin Hood, do not represent mere legends but, in fact, historical persons. Did Arthur defend Britain against the Anglo-Saxons in the Dark Ages of the 6th century? Did Robin thrive as an outlaw in the 100,000-acre royal hunting forest in Nottinghamshire in the 13th century? We will probably never know for sure, but it may have been so, notwithstanding the presence of such fanciful sidekicks as the wizard Merlin and the broad-shouldered Little John.

ARTHUR'S REALM, AND ROBIN'S

Geoffrey of Monmouth's *History of the Kings of England,* an admittedly unreliable text completed circa 1136, leaned on previous histories as it painted a picture of Arthur as a ruler of Roman Britain who defeated the invading Anglo-Saxons and spread his empire to include Ireland, Iceland, Norway and Gaul. Not long after Geoffrey's account appeared, Chrétien de Troyes added the knight Lancelot and his search for the Holy Grail to Arthur's story, and the legend was off and running. It would be embellished in the centuries to come by many—from the poet Sir Thomas Malory to T.H. White to Walt Disney. The more romantic and fanciful Arthur's tale became, the more viciously its underpinnings were disparaged by serious scholars. Wrote archaeologist Nowell Myres: "[N]o figure on the borderline of history and mythology has wasted more of the historian's time."

And yet, there might have been such a man—we simply cannot know for sure. According to Geoffrey's history, his capital city was in Caerleon; it seems to be de Troyes who established Arthur as holding court at "Camelot," in Caerleon. Today, that region is thought to be in Somerset, where a small town of 9,000 souls, called Glastonbury, is seen as the erstwhile perfect kingdom of Camelot. Arthur, many believe, is buried in the ruins of the abbey there.

There is no contemporary place in England named Camelot, but there is certainly a Sherwood Forest, though it is not nearly as large as it was in the 1200s. The road from London to York went through it back then, and so the wood was attractive to highwaymen. A survey of old records shows a fugitive "Robert Hod" in the area in 1226, and still at large the next year as "Robinhud." Did he take from the rich and give to the poor? Well, that we will never know.

CORBIS

THE IMAGE WORKS

There is an unholy fascination with places where people were sent to their deaths in large numbers. "The killing fields" is a term that resonates worldwide, as does "the death camps." Places of extermination were not new to the 20th century, not by a long, woeful shot. Before the Khmer Rouge and the Nazis added their despicable chapters to this awful history, the "advanced" Romans and "sophisticated" Britons had already done their best—which is to say, their worst.

WHERE DWELL THE ONCE DOOMED

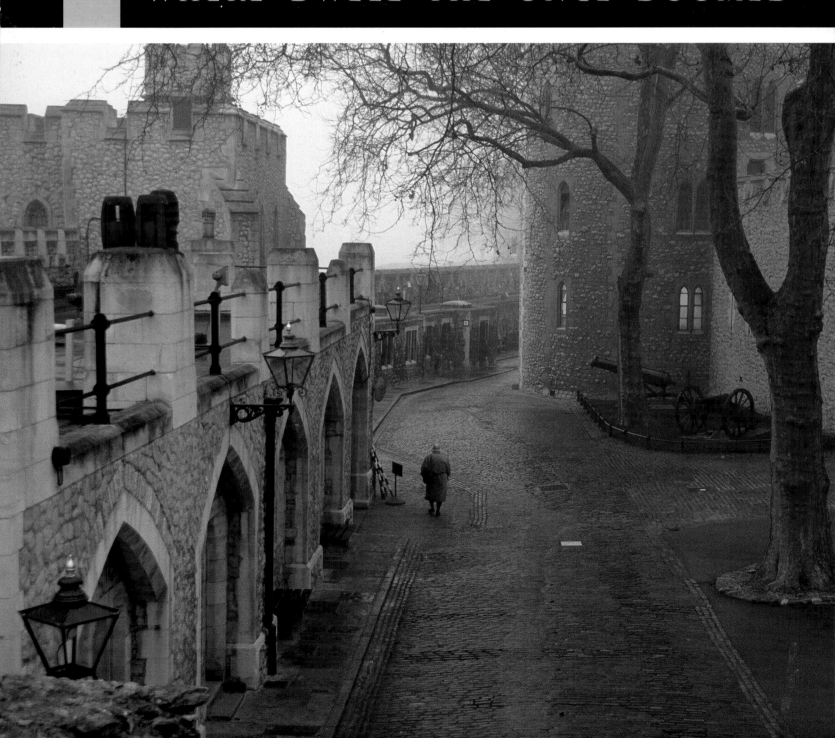

Rome's Colosseum, whose ruins today constitute one of the most awesome of the world's arenas, was first named the Flavian Amphitheatre, circa A.D. 72, by the emperor Vespasian. It was, from the outset, epic. Built on the site of a villa that had once belonged to Nero (a nearby 100-foot statue he had erected of himself, known as a colossus, led to the Flavian's popular name, the Colosseum), its foundation was 42 feet deep and the walls rose 159 feet around an ellipse 615 feet long and 510 feet wide. The infrastructure was of brick, concrete and tufa, and the exterior facade was fashioned from blocks of travertine limestone. As the quarries for the travertine were located a dozen miles outside Rome, and since 3.5 million cubic feet of travertine were required, the logistical problems alone were Sisyphean.

But they were as nothing compared with the trials of those who would provide the "entertainment" there. Vespasian's son Titus dedicated the building in the year 80, setting off a 100-day series of events. Early shows were often headlined by clashes between different species of animals, but soon the productions became increasingly outré. Trained fighters would battle to the death, or occasionally be permitted to live if the warrior had done something special that pleased the crowd. Rome being a pagan state, Christians along with slaves and criminals were fodder for gladiators.

Fodder at the Tower of London, on the banks of the River Thames, often included not just the peasantry but royalty—yesterday's hero, today's condemned. The hundreds of people executed at the Tower for a thousand years—most but not all of them on the chopping block—included petty thieves and three queens, two of them unlucky wives of Henry VIII. The last person to be put to death there, a German spy, was shot by firing squad in 1941. Today, the Tower is a tourist attraction.

Possibly needless to add: It is stated with conviction that both the Colosseum and the Tower are quite thoroughly haunted.

FERDINANDO SCIANNA/MAGNUM

JONATHAN BLAIR/NATIONAL GEOGRAPHIC

● On a typically foggy day in old London town, hardly a soul is stirring near round Lanthorn Tower (left), one of 21 towers (and the second largest) in the Tower of London castle complex. Lanthorn was built in the 13th century during the reign of Henry III, and has served alternately through the years as a palace and a prison—and is now a museum. Above: Rome's Colosseum remains awe-inspiring.

D

Dracula, "The Prince of Darkness," was conceived in London, of all places, in the mind of 19th-century Irish novelist Bram Stoker. But the count's likely progenitor, Prince Vlad Tepes (who also possessed a pro-wrestling–style nickname, "The Impaler"), was born in the land called Transylvania—to considerable circumstance in the mid-1400s.

HORROR STORIES

Truth be told, there's nothing nightmarish about the stunning countryside in this large and legendary province in the middle of Romania. The splendid scenery could as easily have been the setting for *The Sound of Music* as for a horror movie. You can just picture Julie Andrews running through these sunny alpine meadows, singing her merry heart out.

But instead you picture Bela Lugosi as a sharp-fanged demon, operating in the black of night, preying upon innocents. The music, if you conjure any, is full of ominous chords played on a pipe organ.

This is because it is also true that in the distant past a lot of blood was spilled in unspeakable ways in Transylvania. Vlad Tepes was born circa 1431, probably in Sighisoara, which—with its walled citadel perched on a hilltop, its secret gateways to back alleys, its 14th-century clock tower—is one of the most charming examples of a medieval city left in Europe. Outside Brasov, another lovely old city, is the intimidating Bran Castle, where Vlad reigned, forcefully, over Wallachia from 1456 to 1462. These were tempestuous times, and Vlad's father, a former Wallachian prince, had already been assassinated and his brother tortured and buried alive before Vlad ascended to power. And so he had issues. Impalement on stakes in the town square quickly became his signature measure of revenge against the family's enemies, and then continued as the standard punishment for everything from murder down to lying. In 1476, Vlad, having been driven from power, was himself assassinated.

The darkness that surrounds the Dracula legend is certainly substantial, but no more so than that which underlies the famously gorgeous City of Light. Paris sits atop a honeycomb of underground tunnels and passageways—185 miles of them—that in the 18th and 19th centuries were a refuge for robbers, smugglers and fugitives (including the fictive *Phantom of the Opera*). They also served as an adjunct to the city's overstocked cemeteries. Called the Catacombs, this subterranean world still contains the bones of several million Parisians. You can visit the tunnels as a tourist. If you dare.

● Opposite: In 1923, a woman visits a grave below Bran Castle, which at the time was a royal residence within the kingdom of Romania. It is today operated as a museum. Right: Remains of the deceased in Paris's Catacombs, circa 1935. This eerie place, too, is open to the public.

E.O. HOPPÉ/CORBIS

THE IMAGE WORKS

Scattered throughout the Northeast are scores of "mystery caves" and "sacramental sites" that seem to imply that others from the continent of Europe preceded the Vikings to North America. In North Salem, New York, the argument is made that a giant boulder simply could not have been placed so carefully upon its standing stones by glacial action; it must have been put there by man. In the Massachusetts towns of Danvers, Upton and Hopkinton, Celtic-type rock huts called "beehives" have been discovered. And up the road in Salem, New Hampshire, the Mystery Hill Caves are another series of beehives that certainly do resemble those of the Fir Bolg and the Tuatha De Danaan Celts of Ireland.

NEW ENGLAND NOCTURNE

Mystery Hill is sometimes called "America's Stonehenge," and this 30-acre conglomeration of low walls, caves and tunnels on a hillside about 25 miles inland from the Atlantic coast is, if possible, even more enigmatic than England's famous stone circle. Mystery Hill is something of a mess, a chaotic collection of constructs, some seeming to have an astronomical orientation, with others, like the so-called "Sacrificial Stone," appearing to have been built for utilitarian purposes. What might have been sacrificed upon this slab, animals or humans?

More to the point: Who built Mystery Hill? Candidates include not only the Celts but descendants of the Greek and Phoenician cultures of the Mediterranean as well as other early inhabitants—but we may never know.

It was definitely human beings who were sacrificed—or at least put to death—in yet a third town called Salem (the famous one in Massachusetts) in the late 17th century. The Puritans had come to the Massachusetts Bay Colony from England as persecuted men and women; now they became persecutors. Suspicion and unwarranted terror were rife, and civil magistrates reacted irrationally, sending scores of "witches" to prison. Of 200 arrested in 1692 in Salem, 17 died in the village's mean jails. On Gallows Hill, 19 were hanged, then thrown into a burial pit. In 1957, in an era when witch hunts were again a topic of discussion, the Massachusetts legislature posthumously exonerated some of the so-called witches of Salem. As recently as 2001, descendants of five women executed for witchcraft 300 years earlier were again before a Massachusetts tribunal, pressuring the lawmakers to proclaim the women's innocence by name. The Salem witchcraft trials will never die.

According to some who visit Gallows Hill by moonlight, neither will the witches.

AMBER GUILBEAULT

NINA LEEN/LIFE (2)

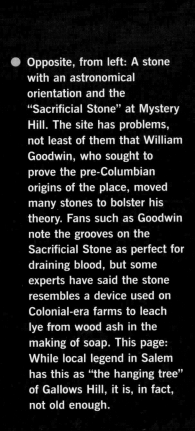

● Opposite, from left: A stone with an astronomical orientation and the "Sacrificial Stone" at Mystery Hill. The site has problems, not least of them that William Goodwin, who sought to prove the pre-Columbian origins of the place, moved many stones to bolster his theory. Fans such as Goodwin note the grooves on the Sacrificial Stone as perfect for draining blood, but some experts have said the stone resembles a device used on Colonial-era farms to leach lye from wood ash in the making of soap. This page: While local legend in Salem has this as "the hanging tree" of Gallows Hill, it is, in fact, not old enough.

The way to consider the eternal strangeness of outer space is through a question: What do we not yet know? We learn more all the time via telescopes, probes, investigations and good old intuitive thinking. Everything we learn changes everything we thought we knew yesterday. We build, slowly, a pile of intelligence—things that we know for sure. But then we gaze into deep space and realize: We have but a small hill of certainty, and the universe is an Everest.

Within a century Pluto is discovered, becomes a planet, then is not a planet again. Consider what "modern man" has thought true (or false) about our own nearby and seemingly knowable satellite, the moon. In 1835, the astronomer John Herschel built a large telescope in South Africa to take advantage of the clear air there. Seizing upon this, pranksters wrote a series of articles in the *New York Sun* that recounted Herschel's discoveries as set forth in the *Edinburgh Journal of Science* (which was, in fact, defunct). Herschel's super telescope revealed lunar rivers and animals—a beaver with no tail, a bison with skin eye-flaps to protect its vision from sunlight. The series' climax detailed furred, flying humans. When told of the worldwide clamor the hoax had caused, Herschel was at first mild: "It is too bad my real discoveries here won't be that exciting." But when folks wouldn't let go, he finally railed, "I have been pestered from all quarters with that ridiculous hoax about the Moon—in English, French, Italian and German!"

In our own day, there are those who persist in believing that man has never set foot on the moon, and that it was all a NASA-orchestrated studio drama. Nothing supports mystery and conjecture quite like outer space, a place most of us cannot visit, cannot quite touch.

Quasars, black holes, wormholes, dark energy, dark matter (over 90 percent of the universe is invisible!): Are we any more certain today about what's happening out there than we were when Jules Verne was sketching deep space for us?

Well, sure we are. Of course we are.

But are we positive?

Well, certainly not.

The Big Bang is being constantly reevaluated. The speed at which matter is rushing away from the center is not what we recently theorized it should be. What does that mean?

What, finally, is the fate of the universe?

No one knows.

No one knows what's out there for us.

NASA

You are looking at a part of the Cassiopeia constellation called W5. It is 7,000 light years away. How big is this region? Fifty light years across. And what is going on there? Stars are being born. The towering pink pillars of cool gas and dust, illuminated at their tips with light from warm, embryonic stars, have been dubbed "Mountains of Creation" by scientists working at NASA's Spitzer Space Telescope observatory. In this image, hundreds of stars, rendered in whites and yellows, are seen by man for the very first time, while the blue stars are a bit older. Scientists believe these star clusters were triggered into existence by radiation and winds from an "initiator" star more than 10 times the mass of our sun. This star is not pictured, but the fingerlike pillars point toward its location above the image frame. We gape at the void and are amazed.

STRANGE BUT NOT TRUE

There just may have been a King Arthur in long-ago Britain, and so there might have been a Camelot (whether it went by that name or some other). And it might have been in or near Glastonbury. Why not?

Vlad Tepes was surely not a vampire but he was evil incarnate, a sadistic killer whose personal history in Transylvania informed the character Dracula. And so a trip to his Bran Castle is worth taking.

None other than Plato wrote of Atlantis, and Plato may well have known of the volcanic eruption that sank most of the Aegean island of Thera, and therefore he may have modeled Atlantis on Thera. So: The remnants of Thera, those ring islands now called Santorini, are something to look at when we consider Atlantis.

Yes, admittedly, as with much else in this book, there may be apocryphal elements to these associations. But serious people tell us that they are deserving of contemplation.

On the other hand, there are many "places" for which there is simply no evidence, nor even clues, in the historic or archaeological record. They are, when you take a hard look at them, about as real as Middle Earth (which is not, by the way, located in New Zealand).

Atlantis and Camelot are hardly alone among paradises that have been lost. Atlantis's rumored Pacific Ocean cousin was an enormous continent, much bigger than Australia, called Mu. It, too, was home to an advanced civilization and it, too, sank beneath the waves eons ago. The story of Mu was first advanced in the late 19th century by a French antiquarian, Augustus Le Plongeon, and then popularized in the 1920s and '30s through a series of books by Le Plongeon's friend, the Anglo-American explorer James Churchward. Churchward started straightforwardly enough with *The Lost Continent of Mu,* but by volume three he was dealing with *The Sacred Symbols of Mu* and by four with *The Cosmic Forces of Mu.* You get the picture. Mu has, ever since, had a New Agey feel. Its symbol, which stands for the Primary Forces of Life radiating outward, was appropriated by the band Led Zeppelin for its fourth album. For all the ink spilled about Mu, no evidence of it has ever been found.

The same is true of Lemuria, yet another lost land situated in

Agartha, the city in the earth's core, is usually accessed through entrances at either the South or North Poles, according to most Agartha experts. The Agarthans may well look like aliens (opposite, top), at least according to Richard Shaver. In the 1940s, he created a sensation when he published a story about his meeting with the remnants of their underground civilization, made up of an Elder Race that had come to earth from another solar system in ancient times, then had largely evacuated via flying saucers. As for Mu: What more to say? It got sunk (right). Music lovers can be thankful for both, however. Not only did a Mu-inspired symbol appear on a Led Zep album, but a terrific late-period Miles Davis double-disc is the funky *Agartha.*

THE IMAGE WORKS

THE EARTHS CORE

Is the earth hollow? Many theorists claim that an opening to an inner world exists at the North Pole. They base their theory on unusual speeds made by polar explorers, auroral phenomenon, weird magnetic effects, and other factors. (See page 144 for complete details.)

HIP/ART RESOURCE, NY

● Sir Walter Raleigh, searching for El Dorado, felt he was close when he described the city of Manoa on an island in a lake called Parimá (the 1599 map above is based on Raleigh's report). The very existence of the lake, never mind Manoa or El Dorado, was later disproved. At right, we have an 1884 painting by E.V. Luminais, "Flight of King Gradlon." According to a legend regarding the city of Ys, Gradlon, King of Cornouaille, was beseeched by his ocean-loving daughter, Dahut, to build a city below sea level. Silly girl. Sillier still when she, as ruler of Ys, made it a northern Sodom with her willful orgies, then was seduced by the devil who, once inside the city, unleashed a flood. In the painting, Gradlon sacrifices his daughter to the onrushing sea. Just deserts.

either the Indian or Pacific Ocean. Same is true of Thule, the missing island in the Scandinavian belt. Same is true of Agartha, the city said to be situated in—get this!—the earth's core.

There are many variations on the Hollow Earth or "underground civilizations" theory, Agartha being at the heart of several of them. The people that live there are very advanced, of course, and may or may not be directly related to Lemurians, Atlanteans and/or the citizens of Mu.

That our planet might be hollow, and that there might be denizens at its core, proved an entertaining idea in the fictions of those such as Jules Verne, Edgar Allan Poe and Edgar Rice Burroughs. But scientists tell us the earth has a solid mantle and core, and that no one lives there. Still Agartha has adherents, and always will.

There are myths regarding cities on the surface, too. Ys was an urban center on the coast of France that, like New Orleans, was built largely below sea level. A dam protected Ys until the devil himself caused a flood that sank the city forevermore.

Among cities of fabulous riches whose existence has never been proven we have Quivira and Cibola, two of the legendary Seven Cities of Gold that were sought by conquistadors throughout the Americas; we have the City of the Caesars, which was nestled between a mountain made of gold and another of diamonds in Patagonia; and we have, particularly, El Dorado, the ne plus ultra never-was city of gold in South America. El Dorado (Spanish for "the gilded one") was quested after

EVARISTE VITAL LUMINAIS/GETTY

by adventurers of such considerable renown as Francisco Pizarro (in 1541) and Sir Walter Raleigh (in 1595), always to no avail. Although one explorer claimed to have seen the city, he never took that claim to the bank.

Places that do not exist and never did are not exclusive to our planet. Phaeton was supposed to be a deceased planet that once orbited our sun

between Mars and Jupiter. It was first posited in the early 19th century by the astronomer Heinrich Wilhelm Matthäus Olbers, a man of intelligence and renown. The theory went that Phaeton perhaps ventured too close to Jupiter and was destroyed by that enormous planet's fierce gravity, or that Phaeton was smashed by an asteroid or rogue planet, or that some horrific internal combustion finished it off. Today's scientists disregard all Phaeton conjectures, although the idea that the solar system is missing one of its original members persists in some pseudoscientific corners.

All of these ideas persist in certain corners—in nooks and crannies populated by true believers. What their belief proves above all else: The power of myth is strong beyond measure.

Driftwood Library
of Lincoln City
801 S.W. Highway 101, #201
Lincoln City, Oregon 97367

CORBIS

On the night before Halloween in 1938, William Dock, 76, of Grovers Mill, New Jersey, stands ready with his trusty shotgun to ward off Martians. Why does he do this? Because on this night, Orson Welles, in dramatizing H.G. Wells's novel *The War of the Worlds* on his Mercury Theater of the Air radio broadcast, has told America that the first spaceship of the Martian invasion has landed in Grovers Mill. People nationwide are terrified, and many locals (not including the intrepid William Dock) are fleeing the area.

Where is the strangeness and the truth in this story? Martians did not invade (not in 1938, anyway). But real people really did flee, and Mr. Dock really did ready his gun.

There is no profit in sitting in judgment on the human condition. It is a strange world we live in, and strange things will continue to happen tomorrow and the next day. The truth is, we are all frail— frail when dealing with harsh certainties, and even more frail when we open our minds to wonder.